Synthese Library

Studies in Epistemology, Logic, Methodology, and Philosophy of Science

Volume 490

The aim of *Synthese Library* is to provide a forum for the best current work in the methodology and philosophy of science and in epistemology, all broadly understood. A wide variety of different approaches have traditionally been represented in the Library, and every effort is made to maintain this variety, not for its own sake, but because we believe that there are many fruitful and illuminating approaches to the philosophy of science and related disciplines.

Special attention is paid to methodological studies which illustrate the interplay of empirical and philosophical viewpoints and to contributions to the formal (logical, set-theoretical, mathematical, information-theoretical, decision-theoretical, etc.) methodology of empirical sciences. Likewise, the applications of logical methods to epistemology as well as philosophically and methodologically relevant studies in logic are strongly encouraged. The emphasis on logic will be tempered by interest in the psychological, historical, and sociological aspects of science. In addition to monographs *Synthese Library* publishes thematically unified anthologies and edited volumes with a well-defined topical focus inside the aim and scope of the book series. The contributions in the volumes are expected to be focused and structurally organized in accordance with the central theme(s), and should be tied together by an extensive editorial introduction or set of introductions if the volume is divided into parts. An extensive bibliography and index are mandatory.

Massimiliano Carrara • Enrico Martino

Arbitrary Reference in Logic and Mathematics

Springer

Massimiliano Carrara
FISPPA Department
Padua University
Padova, Italy

Enrico Martino
FISPPA Department
Padua University
Padova, Italy

ISSN 0166-6991 ISSN 2542-8292 (electronic)
Synthese Library
ISBN 978-3-031-66451-9 ISBN 978-3-031-66452-6 (eBook)
https://doi.org/10.1007/978-3-031-66452-6

This Springer imprint is published by the registered company Springer Nature Switzerland AG
The registered company address is: Gewerbestrasse 11, 6330 Cham, Switzerland

Preface

The purpose of this book is to show how arbitrary reference can throw new light to the foundations of logic and mathematics.

Working mathematicians refer very often, in their reasoning, to an arbitrary object of the universe of discourse. They employ expressions such as "Let a be an arbitrary set" and reason about a as if it were a well-determined set. We have called such a locution a "magic locution", because it puts the puzzling questions:

(Q1) How can one fix the reference of a?
(Q2) How can *any* set be capable of being selected among the enormous infinity of sets?

Questions like (Q1) and (Q2) are usually neglected by the working mathematician, because the means of fixing the reference of a does not enter in her reasoning. All she needs is to *imagine* that reference has been *somehow* fixed. That's right, but this very consideration suggests the essential role of *imagination* in mathematical reasoning and the opportunity of making it more perspicuous. Following this suggestion, we introduce an ideal agent, free of empirical limitations, who has access to any individual of the domain of quantification. We propose a revision and a development of J.S. Mill's and P. Kitcher's attempts of interpreting the mathematical language as concerning ideal actions instead of abstract mathematical entities. Subsequently we extend the singular agent to an ideal team of infinitely many agents, as many as the individuals of a first order domain of quantification. The job of such agents is that of performing each of them, ad libitum, a simultaneous act of choice of an individual. In this way, singular arbitrary reference is generalized to plural arbitrary reference.

By using this device, we defend G. Boolos' controversial thesis of the ontological innocence of second-order logic. Boolos' argument, based on his reinterpretation of second-order logic in terms of *plural quantification*, has faced a criticism of several authors, moved from the suspicion that plural quantification would surreptitiously exploit a hidden notion of set. We will argue, however, that such a suspicion can be superseded by explaining plural quantification in terms of *plural arbitrary reference* with a suitable semantics for plural choices.

Using our semantics, we analyze and revise Lewis' work on *megethology*, a reconstruction of set theory based on the combination of plural quantification and mereology. In particular, we discuss Lewis' thesis of the ontological innocence of mereology. We argue that the notion of mereological fusion, so as usually understood, is committal to larger and larger infinities, provided a countable infinity of individuals is given. Thereof we maintain that *usual* mereology is far from being ontological innocent. We show, however, how a virtual mereology (**VM**) can be constructed, where the role of fusions is played by suitable plural acts of choice, so that no new entity is generated by the operation of fusion.

We conclude that the ideal team of agents is a powerful source of new insight on the foundations of logic and mathematics.

This book collects and elaborates some ideas found in these papers:

M. Carrara and E. Martino. A Notion of Logical Concept Based on Plural Reference. *Acta Analytica*, 33: 19–33, **2018** (DOI: 10.1007/s12136-017-0327-2).

M. Carrara and E. Martino. The Mereological Foundation of Megethology. *Journal of Philosophical Logic* 45: 227–235, **2016** (DOI: 10.1007/s10992-015-9373-7).

M. Carrara and E Martino. Grounding Megethology on Plural Reference. *Studia Logica* 103: 697–711, **2015** (DOI: 10.1007/s11225-014-9585-9).

M. Carrara and E. Martino. To Be is to Be the Object of a Possible Act of Choice. *Studia Logica* 96: 289–313, **2010** (DOI: 10.1007/s11225-010-9282-2).

Arbitrary reference in logic and mathematics is the result of a rather lengthy collaboration that we began in late 2006. A considerable amount of time has passed since then, although we have never ceased to discuss of arbitrary reference, mereology, and other issues in logic and philosophy of logic, such as the notion of proof and that of negation, which have found only a bit of space in this book.

It narrates a bit of our *research journey* undertaken together over the years; perhaps this is also why it took us so long to finish it! We liked the idea of continuing to discuss these matters, continuing our research journey.

It is for this reason that we chose to conclude the writing of this book with a poem, "Ithaca," that the poet Konstantinos Petrou Kavafis (translated by E. Keely) dedicated to the journey. Kavafis in this poem reminds us not to be in a hurry to reach our destination. These are such beautiful verses that we like to share them with the readers of this book (or more simply the readers of this preface).

Ithaca (by Konstantinos P. Kavafis)[1]

As you set out for Ithaka
hope your road is a long one,
full of adventure, full of discovery.
Laistrygonians, Cyclops,
angry Poseidon—don't be afraid of them:
you'll never find things like that on your way

[1] C. P. Cavafy, *Collected Poems*. Translated by Edmund Keeley and Philip Sherrard. Princeton University Press 1975.

as long as you keep your thoughts raised high,
as long as a rare excitement
stirs your spirit and your body.
Laistrygonians, Cyclops,
wild Poseidon—you won't encounter them
unless you bring them along inside your soul,
unless your soul sets them up in front of you.

Hope your road is a long one.
May there be many summer mornings when,
with what pleasure, what joy,
you enter harbors you're seeing for the first time;
may you stop at Phoenician trading stations
to buy fine things,
mother of pearl and coral, amber and ebony,
sensual perfume of every kind—
as many sensual perfumes as you can;
and may you visit many Egyptian cities
to learn and go on learning from their scholars.

Keep Ithaka always in your mind.
Arriving there is what you're destined for.
But don't hurry the journey at all.
Better if it lasts for years,
so you're old by the time you reach the island,
wealthy with all you've gained on the way,
not expecting Ithaka to make you rich.

Ithaka gave you the marvelous journey.
Without her you wouldn't have set out.
She has nothing left to give you now.

And if you find her poor, Ithaka won't have fooled you.
Wise as you will have become, so full of experience,
you'll have understood by then what these Ithakas mean.

At the end of this preface, there can be no shortage of acknowledgments. We want to give plural thanks to all those who have read parts of this work, in earlier versions, suggesting revisions, cuts, and improvements. We also want to thank anonymous reviewers who read draft versions of the book for their detailed comments, suggestions, and improvements. We hope that this revised version of the book will satisfy their requests. Finally, we want to express our gratitude to Otávio Bueno, Editor-in-Chief of the *Synthese Library series. Studies in Epistemology, Logic, Methodology, and Philosophy of Science*, for supporting with patience (!) the editorial project and for wanting it to be published in the series he directed, and to Christopher Wilby Associate Editor of Springer for the editorial work.

This book is dedicated to our beloved Elena and Livia. And it is also for the beloved son of Massimiliano, Pietro in the hope that he may see the Otaru Harbor dock again.

Padova, Italy Massimiliano Carrara
4 April 2024 Enrico Martino

Contents

Chapter 1
On Arbitrary Reference

Abstract In this chapter we introduce (**PAR**), the *Principle of Arbitrary Reference*. According to **PAR** any object of the universe of discourse is capable of been picked out by an act of arbitrary reference. We argue that **PAR** is essential for both formal and informal logical deduction, as well as for the semantics of quantifiers. We propose to understand *arbitrary reference* as direct reference via an ideal act of choice, setting the stage for further developments in later chapters.

1.1 Introduction

Mathematicians use very often in their reasoning the following:

Magic locution (**ML**) "let a be an arbitrary object of the universe of discourse".

The magic feature of **ML** consists in the fact that, in general, the universe of discourse U is uncountably infinite and no means is available for referring to any member of U. For instance, only countably many real numbers are describable in a countable language, so that most of the real numbers are undescribable. However, introducing a with **ML**, we pretend that the letter "a" picks out a well-determined–though unknown–real number and reason as if any real number might be the one picked out.

It seems therefore, at least prima facie, that mathematical reasoning exploits the following:

Principle of arbitrary reference (**PAR**). Any object of the universe of discourse is capable of been picked out by an act of reference.

M. Carrara, E. Martino, *Arbitrary Reference in Logic and Mathematics*, Synthese Library 490, https://doi.org/10.1007/978-3-031-66452-6_1

Question: Is **PAR** an essential part of logical reasoning?

Aim of this chapter is to give a positive answer to the above question. We argue that **PAR** is essential for formal and informal logical deduction.[1] Moreover, we argue that **PAR** plays an essential role even in the semantics of quantifiers. We sketch a picture for understanding arbitrary reference as a sort of direct reference realized by means of an *ideal act of choice*. Here we just sketch our proposal. A complete picture will be fully developed in the next chapters.

1.2 On PAR

Consider our *magic locution* **ML**:

(**ML**) "let *a* be an arbitrary object of the universe of discourse".

One may think that **ML** is nothing but a way of speaking, used in informal mathematical reasoning, and that **PAR** is not essential for logical deduction. So, for example, one may argue that reasoning about an arbitrary triangle, does not presuppose the possibility of singling out any triangle, since the reasoning does not exploit the identity of the triangle in question. Speaking of an arbitrary triangle, the argument goes, amounts to speaking of *all* triangles *simultaneously* and not about a *well-determined* triangle.

Mutatis mutandis it is the same train of thoughts adopted by Ramsey in (1925) when he criticizes Russell's argument that an infinite class, since its members cannot be listed, must be describable by a propositional function. Ramsey maintains that the argument rests on the belief that if one cannot speak of an object individually one cannot speak of it at all. According to him such belief is erroneous, since one can speak of all objects by means of quantified propositions.

However, observe that such criticism is deceptive, because it overlooks the fact that *arbitrary reference* is essential for establishing quantified statements. This fact was clearly emphasized by Russell (in this long passage):

> The general enunciation tells us something about (say) all triangles, while the particular enunciation takes one triangle and asserts the same thing of this one triangle. But the triangle taken is any triangle, not some one special triangle; and thus, although, throughout the proof, only one triangle is dealt with, yet the proof retains its generality. If we say: "Let ABC be a triangle, then the sides AB and AC are together greater than the side BC", we are saying something about one triangle, not about all triangles; but the one triangle concerned is absolutely ambiguous, and our statement consequently is also absolutely ambiguous. We

[1] The fact that the Principle of Arbitrary Reference (**PAR**) is essential for logical deduction is not a novelty. Indeed, the presence of **PAR** is already discernible in the work of Euclid, illustrating its deep-rooted significance in the foundations of logic and mathematics. For a general discussion on Euclidean geometry and its impact on the development of mathematical thought, offering context for Euclid's work within the broader history of mathematics see Boyer and Merzbach (2011); see also Martin (2012) where the logical underpinnings of geometry, related to concepts of arbitrariness and generality are examined.

> do not affirm any one definite proposition, but an undetermined one of all the propositions resulting from supposing ABC to be this or that triangle. This notion of ambiguous assertion is very important, and it is vital not to confound an ambiguous assertion with the definite assertion that the same thing holds in all cases. The distinction between (1) asserting any value of a propositional function and (2) asserting that the function is always true is present throughout mathematics, as it is in Euclid's distinction of general and particular enunciations. In any chain of mathematical reasoning, the objects whose properties are being investigated are the arguments to any value of some propositional function.... For this reason, when any value of a propositional function is asserted, the argument...is called a real variable, whereas, when a function is said to be always true, or to be not always true, the argument is called an apparent variable. ... If fx is a propositional function, we will denote by "$(x).fx$" the proposition "fx is always true" ....Then the distinction between the assertion of all values and the assertion of any is the distinction between (1) asserting $(x).fx$ and (2) asserting fx where x is undetermined. The latter differs from the former in that it cannot be treated as one determinate proposition. The distinction between asserting fx and asserting $(x).fx$ was, I believe, first emphasized by Frege [...]. His reason for introducing the distinction explicitly was the same which had caused it to be present in the practice of mathematicians, namely, that deduction can only be effected with real variables, not with apparent variables. In the case of Euclid's proofs, this is evident: we need (say) some one triangle ABC to reason about, though it does not matter what triangle it is. The triangle ABC is a real variable; and although it is any triangle, it remains the same triangle throughout the argument. But in the general enunciation the triangle is an apparent variable. If we adhere to the apparent variable, we cannot perform any deduction, and this is why in all proofs real variables have to be used. (Russell, 1908, pp. 156–157)

The above passage clearly enlightens the need of reasoning about a *single arbitrary object* in order to reach a universal conclusion.

However, we are not entirely in agreement with Russell's explanation of the nature of arbitrary reference. To say that any triangle is not some one special triangle, but that it is absolutely ambiguous, might erroneously suggest that the triangle we are concerned with is a strange object enjoying the odd property of *being absolutely ambiguous*. But, of course, an ontology of ambiguous objects would be far from being desirable.

Take, for example, the well-known ontology of *arbitrary objects*, developed by Kit Fine in (1985). In this perspective an arbitrary object is conceived as an abstract object enjoying all properties shared by a certain manifold of individual objects. We find such an ontology *ad hoc* and highly counterintuitive. Anyway, the move of extending the ontology cannot serve the purpose of explaining arbitrary reference, since this notion concerns all kind of objects. How can we arbitrarily refer to an "arbitrary object"?[2]

Alternatively, Russell's explanation might suggest that what is ambiguous is the *act of referring* to any arbitrary triangle, in the sense that it is undetermined which triangle it refers to. This second alternative seems to be held by Frege, who explicitly rejects the first one when he says:

[2] For a similar criticism to *arbitrary objects* see Breckenridge and Magidor (2012).

> (Mr. E. Czuber)... defines a variable as an indefinite number. But are there indefinite numbers? Must numbers be divided into definite and indefinite? Are there indefinite men? Must not every object be definite? "But is not the number n indefinite?" I am not acquainted with the number n. "n" is not the proper name of any number, definite or indefinite. Nevertheless, we do sometimes say "the number n". How is this possible? Such an expression must be considered in a context. Let us take an example. "If the number n is even, then $cos\ n\pi = 1$". Here only the whole has a sense, not the antecedent by itself nor the consequent by itself. The question whether the number n is even cannot be answered; no more can the question whether $cos\ n\pi = 1$. For an answer to be given, "n" would have to be the proper name of a number, and in that case this would necessarily be a definite one. We write the letter n in order to achieve generality. This presupposes that, if we replace it by the name of a number, both antecedent and consequent receive a sense. Of course we may speak of indefiniteness here; but here the word "indefinite" is not an adjective of "number", but "indefinitely" is an adverb, e.g., of the verb "to indicate". We cannot say that "n" designates an indefinite number, but we can say that it indicates numbers indefinitely. (Frege, 1904, p. 287)

But, besides being quite obscure what an act of indicating indefinitely is, such indefiniteness would be incompatible with the essential fact, clearly stressed by Russell, that, in the whole course of the proof, the referent is always the same. This means that the letter "a" indicates the same object in all its occurrences in the contest of the proof. If the referents of such occurrences were not well-determined objects, it would be meaningless to say that they are the same. Perhaps Frege believed this objection to be superseded by his doctrine of functions as unsaturated entities. Indeed, in *Function and Concept* he says:

> People who use the word "function" ordinarily have in mind expressions in which a number is just indicated indefinitely by the letter x, e.g. "$2x^2+x$"... (But) x must not be considered as belonging to the function; this letter only serves to indicate the kind of supplementation that is needed; it enables one to recognize the places where the sign for the argument must go in. (Frege, 1891, p. 141)

It seems that, according to Frege, "to indicate indefinitely" really means "not to indicate anything at all". The variables occurring in a function expression do not denote anything; they are mere placeholders marking the gaps to be filled by individual names. According to this view, a reasoning about an arbitrary object x may be regarded as a *schema of reasoning*, i.e. as a function which maps every object into the reasoning obtained from the schema by replacing "x" with a name of such an object.

It is plain, however, that such a schema can have all the desired instances only if every object is capable of being singled out and named. Thus regarding "x" as a schematic letter is of no help to avoid **PAR**.

We think therefore that the ambiguity shown by Russell is to be understood in a purely *epistemic sense*. Referring to an arbitrary object a amounts to supposing that "a" designates an unknown, though well-determined, object. Being well-determined justifies the behavior of "a", in the course of reasoning, as a name designating the same object in all its occurrences. On the other hand, being unknown guarantees that all that is recognized for it holds also for any other object of the domain.

In today's formal logic Russell's distinction between real and apparent variables is faithfully reproduced, with a sheer change in terminology, by the well-known distinction between free and quantified variables. Singular reference plays an essential role in quantification theory. This fact is made quite perspicuous by the quantification rules in natural deduction. Take the rules for the existential quantifier. According to the elimination rule for the existential quantifier, in order to derive a conclusion A from an existential assumption $\exists x Px$, one has to assume Pb (where b is a fresh name) and derive A from Pb (with the due restrictions). This rule is justifiable only if it is granted that, under the mere existential assumption, we can refer to an object b such that Pb. If it were not granted that every object of the domain is capable of singular reference, there would be no evidence of the soundness of this rule: the formula Px might be satisfied only by some object incapable of singular reference. Similarly for the introduction rule of the universal quantifier. The justification of classical natural deduction rests therefore on the hidden assumption of **PAR**. Thus natural deduction does justice to the informal locution:

"Let a be an object such that Pa".

And since **PAR** is exploited in deductive reasoning, we think that it is to be regarded as a logical principle. Moreover, from the above discussion, it may seem that the role of **PAR** is confined to informal deduction. On the contrary, in the following, we argue that **PAR** plays an essential role even in the semantics of quantifiers.

Consider the usual inductive definition of truth (or satisfaction) for first-order logic. Take, for simplicity, the universal quantification of an atomic formula. If P is a predicate letter, the truth-value of $\forall x Px$ is defined by the clause:

$\forall x Px$ is true if Px is true relative to every assignment of an individual to a variable x.

The definition of the truth-value of $\forall x Px$ presupposes the definition of the truth-value of Px relative to an arbitrary assignment. One can easily realize that **PAR** is masked behind the familiar use of the notion of an assignment, defined as a function from variables to individuals. One can see here a twofold role of **PAR**; it guarantees:

(i) that any object can be the value of some assignment, and
(ii) that one can pick out an arbitrary assignment, understood as a mathematical function.

Demanding to a set-theoretical function the task of referring to an arbitrary object has the mere effect of shifting the problem of arbitrary reference from the object-language to the meta-language. Thus, in order to avoid a regress ad infinitum, we must take, at some stage, arbitrary reference as primitive. In the primitive sense, arbitrary reference is to be thought of as accomplished by an act, not by an entity.

The thesis that an arbitrary object is nothing but an unknown ordinary object is maintained also by R. Price in (1962). The author rightly concludes that natural deduction does not need arbitrary objects understood as special abstract objects. However, he seems also to conclude that the interpretation of "arbitrary" as unknown can avoid any appeal to a principle like **PAR**:

> If someone asks (to a mathematician) to justify a general proof derived from an "arbitrary" triangle, he will point out that he made no use of such factors as made that triangle special. He does not appeal to a Lockeian abstract idea but rather to the scrupulous use he has made of a definite triangle. Of course he might add that he could have used any triangle (or very nearly any) but this follows from the care, which he took... (Italics ours) (Price, 1962, p. 95)

The author seems to overlook the fact that the possibility of using any other triangle is granted just by the possibility of referring to any other triangle. For this reason, the epistemic interpretation of "arbitrary" cannot avoid **PAR**.

1.3 Is PAR a Semantic Principle?

A principle of *arbitrary reference* has been recently defended by W. Breckenridge and O. Magidor in (2012).[3] They express the principle as follows:

(**AR**) It is possible to fix the reference of an expression arbitrarily. When we do so, the expression receives its ordinary kind of semantic value, though we do not and cannot know which value in particular it receives.

They rightly argue that **AR** provides the best explanation of the usual mathematical reasoning. In particular they hold that a stipulation as:

(*) "Let n be an arbitrary number"

fixes a well determined, but unknowable, number as the referent of "n".

They do not provide any explanation, however, of how the link between "n" and its referent is generated. They hold that it is a brute primitive semantic fact that (*) fixes a precise, though unknowable, number.

But what is a semantic fact? If semantics is the theory of meaning, the authors seem to claim that (*) determines a number in virtue of its very meaning. But what is highly problematic is just the meaning of the expression in question, in particular the meaning of the word "arbitrary".

Observe that the mere fact that the locution (*) determines a precise but unknowable number, apart from being mysterious, is inadequate to justify the introduction rule of the universal quantifier (I$\forall$). Suppose that, reasoning about the number n introduced by locution (*), you have proved Pn. Why can you infer $\forall x Px$? Prima facie, the obvious answer is: since all that you know about n is that it is a number, what you have proved for n must hold for every number. But, on reflection, one can deny that your unique piece of information about n is that it is a number; in fact you also know that it is the number determined through locution (*).

[3] For a general discussion see also Hawthorne and Manley (2012).

And this piece of information is essentially exploited in your argument for Pn, since what justifies your treatment of n as a fixed number is just the fact that it has been fixed by that locution. And the mere fact that a certain unknown number has been fixed by a certain locution does not explain the meaning of the word "arbitrarily" occurring in **AR**; so it cannot guarantee that what holds for that number holds for all numbers.

We conclude that the brute fact that (*) introduces a well-determined unknowable number is inadequate to justify the rule of universal introduction (I∀).

1.4 PAR and Game-Theoretical Semantics

The commitment to **PAR** is quite evident in J. Hintikka's game theoretical semantics for first-order logic (see, e.g. Hintikka, 1996). In this semantics the meaning of logical constants is explained in terms of acts of choice. Hintikka associates with every sentence of a first order language a game between two ideal players, the *verifier* and the *falsifier*, who are trying, respectively, to verify and to falsify the sentence. Then, he defines the truth and the falsity of a sentence as the existence of a *winning strategy*, respectively, for the verifier and for the falsifier. The game rules are defined in terms of arbitrary choices of individuals by the players and introduction of names for the chosen individuals. The definition proceeds by induction on the complexity of the sentence. In particular, the clause for the existential quantifier is the following:

A play relative to $\exists x Sx$ starts with a choice of an individual b by the verifier. Then the play continues as for Sb.

The clause relative to the universal quantifier ∀ is similar, with the choice made by the falsifier.

Hintikka observes that the name "b" does not necessarily belong to the given language but that, since the length of a sentence is finite, any play requires only finitely many new names. Hintikka takes for granted the ideal possibility of choosing any individual and giving it a name. He stresses the constructive flavor of his semantics, arising, in his view, from the fact that the meaning of logical constants is grounded on the notion of action. He proves the soundness of classical logic for his semantics and concludes that, in spite of the intuitionist tenets, classical logic is constructively justified. Indeed Hintikka's game rules are perfectly intelligible from an intuitionistic viewpoint. However, his proof of soundness rests on a platonist understanding of a winning strategy: once the game rules have been established, he regards as a well-determined objective fact the existence or non existence of a winning strategy for the verifier or for the falsifier. So Hintikka's proof that for every game there is a winning strategy for one of the two players fails intuitionistically. This explains why Hintikka's semantics turns out to be equivalent to the usual classical Tarskian semantics. Hintikka's "constructivism", based on the action of choosing, has nothing to do with intuitionist antirealism.

We think, however, that Hintikka's philosophical perspective is coherent and that it supplies a faithful analysis of the usual classical mathematical reasoning. In particular, it makes explicit a "constructive" aspect hidden even behind the classical conception of mathematics. It seems to regard the possibility in principle of choosing any individual as implicit in the logical concept of object.

The idea of an *act of choice* seems to provide an appropriate framework for understanding the notion of arbitrary reference. In the following we will pursue the idea that arbitrary reference is a sort of direct reference based on an imaginary act of choice. This same idea will be exploited in details in the next chapters.

1.5 The Principle of Arbitrary Choice (PAC)

Let us imagine that we have direct access to an ideal agent, who in turn has direct access to every object: he can choose any object at will (here we are identifying ourselves with the working mathematician carrying on the mathematical reasoning). We can explain the locution:

"Let a be an arbitrary object"

as follows:

We ask the agent to choose an object at her will without communicating us anything about the chosen object and call it "a".

The fact that we do not know the object a guarantees that what we can know about a must hold for *any* object. That is the reason why we cannot imagine ourselves as playing the role of the ideal agent. It is clear that the adjective "arbitrary" does not concern the nature of the chosen object, but the freedom of the act of choice. Accordingly we will specify **PAR** as a principle of arbitrary choice:

(**PAC**) Every object of the universe of discourse is capable of being chosen by the ideal agent.

This picture can be seen as an idealized version of Kripke's explanation of his notion of *direct reference*:

> A rough statement of the theory might be the following: an initial "baptism" takes place. Here the object may be named by ostension... When the name is "passed from link to link," the receiver of the name must, I think, intend when he learns it to use it with the same reference as the man from whom he heard it. (Kripke, 1980, 96)

Similarly, in our framework, the ideal agent makes a baptism of an object at his will by ostension and we use the name for referring to that (unknown to us) object. In this way we transfer to the mathematical language Kripke's proposal for natural language.

Of course, **PAC** faces immediately the problem of how to choose by ostension a mathematical object.

One can object that, since mathematical objects are abstract, they are not capable of reference by ostension. What does it mean to choose an infinite entity such as a real number or an infinite set? It is hard to give a general answer to this question, since any answer depends essentially on how mathematical entities are conceived. We suggest considering **PAC** as a desirable constraint for a conception of mathematical objects. The structuralist development of mathematics has shown that mathematical theories do not determine the specific nature of the entities they are talking about. Therefore, given any mathematical theory, the possibility is left of searching for models built up from objects capable of direct reference. For first-order arithmetic a suitable model is Hilbert's model of numerals thought of as finite strings of strokes. These are, in C. Parsons' words, *quasi-concrete objects*, i.e. types of spatio-temporal objects, the accessibility to which requires only a minimal idealization of the agent, needed for dealing with the mathematical infinite (Parsons, 1990). All that the agent is expected to be able to do is to write down any finite (however long) string of strokes.

In the sequel, it will be evident that a countable infinity of concrete objects is sufficient even for a reconstruction of the classical continuum. For the moment observe that, at least from a logical point of view, Hilbert's strings are sufficient for interpreting any mathematical theory. For, as is well known, it follows from the Loewenheim-Skolem theorem that every first order consistent theory has a countable model. Although such a model is, in general, far from being the intended model of the given theory (say set theory), it is a relevant fact that any consistent first order theory can be interpreted in domains of quasi-concrete objects and that these interpretations are sufficient to guarantee the completeness theorem.

Observe that our ideal agent, unlike the Brouwerian idealized mathematician, has no other job than that of performing arbitrary acts of choice. He is not expected to have the capacity of restricting his choices to objects satisfying some required condition. The inferential step from $\exists x Fx$ to "let a be an arbitrary object such that Fa" is justified by referring to a completely arbitrary choice of the agent, calling "a" the chosen object and assuming Fa. Though this assumption may be false, any logical consequence B we can draw from it must be true under the hypothesis $\exists x Fx$ (provided that a does not occur in B). For, the existential hypothesis and **PAC** assure that the agent could have chosen (though unconsciously) an object satisfying F; and since we don't know anything about a, even in that (possibly counterfactual) case, our reasoning would be correct. And since the proviso on B assures that the truth-value of B is independent of the effective choice of the agent, B turns out to be true under the mere existential assumption.

1.6 PAC and the Axiom of Choice

Let us compare **PAC** with the celebrated set-theoretical *axiom of choice*. **PAC** does not assert the existence of any mathematical object; it explains the meaning of free variables and justifies their inferential role. In contrast, the axiom of choice states that, given any set α of non empty pairwise disjoint sets, there is a set β (call it the choice set) sharing a unique element with each member of α.

From a logical point of view, no act of choice is involved in the understanding of the content of this axiom (besides that, with which we are concerned, implicit in the general concept of an object). It is a purely existential statement expressible in the language of set theory: as all sets, a choice set exists, in a realistic perspective, quite independently of any human action.

On the other hand, the act of choice seems to constitute the intuitive ground for the existence of a choice set. It is usually agreed that the axiom serves the purpose of granting the existence of a choice set even when its elements are not singled out by any propositional function. The existence of such a set seems to be intuitively justified by thinking of its members as arbitrarily chosen. This aspect was just the main source of the well-known dogged opposition to the axiom: it was charged of introducing into mathematics indefinable sets (a set being definable if it is the extension of a propositional function without parameters). It may be puzzling that, if α is finite, the axiom of choice is not needed for the existence of β, even when it is indefinable. In particular, given any single non empty set α, no axiom of choice is needed in order to guarantee the existence of a subset β of α with a unique member. The reason is to be found in a hidden application of **PAC**, implicit in quantification theory.

Briefly, the proof runs as follows: let a be a member of α; by the pairing axiom there is a set β whose unique member is a. It is clear that this kind of reasoning (formalizable in axiomatic set theory) is correct only under the hidden assumption that any object of the universe of sets is capable of being chosen.

The role of choice in mathematics, contrary to a widespread belief, is far from being restricted to the use of the axiom of choice; it is pervasive of the whole mathematics and logic. The axiom of choice seems to exploit the idea of choice in a more problematic way, since it involves the possibility of a simultaneous choice of infinitely many objects. In the next chapter we will argue, however, that this possibility is already implicit in the usual notion of a set as constituted by its members (in contrast with the logical notion of a class as extension of a property).

One may object, against the need of **PAC**, that the mathematical language can be understood by direct extrapolation from the ordinary talk about concrete objects (for which reference is not problematic), so that, in particular, arbitrary reference to mathematical objects would be immediately intelligible without any need of further explanations. According to this objection, the familiar understanding of a talk about any man, any horse or any pencil would make a talk about any real number immediately meaningful. This opinion seems to be shared by Shapiro. He notes the elusiveness of reference in mathematics, but doesn't seem to find it very problematic. Consider the following long passage:

> Probably the most baffling, and intriguing, semantic notion is that of reference. The underlying philosophical issue is sometimes called the " 'fido'-fido problem". How does a term come to denote a particular object? What is the nature of the relationship between a singular term ("Fido") and the object that it denotes (Fido), if it denotes anything? Notice that model theory, by itself, has virtually nothing to say on this issue. In textbook

> developments of model theory, reference is taken as an unexplicated primitive. It is simply stipulated that an "interpretation" includes a function from the individual constants to the domain of discourse. This is a mere shell of the reference relation.
> [...] As far as the model-theoretic scheme goes, it does not matter how this reference is to be accomplished or whether it can be accomplished in accordance with some theory or other. There is nothing problematic in the abstract consideration of models whose domains are beyond all causal contact. As far as model theory goes, reference can be any function between the singular terms of the language and the ontology[...]
> It is fair to say that when it comes to mathematics and theories of other abstracta, realism in ontology often falters over reference (about as much as it falters over epistemology). If we assume that ordinary languages are understood and if we accept the premise that model theory captures the structure of ordinary interpreted languages, then we can do better. There is, of course, no consensus on how reference to ordinary physical objects is accomplished. The theories are legion. I do presume, however, that reference to proverbial medium-sized physical objects is accomplished... .Understanding how to use ordinary language involves an understanding, at some level, of reference... (however it works). (Shapiro, 1997, p. 139)

In the above long quotation Shapiro seems to hold that the comprehension of the notion of reference, acquired from the use of natural language, is sufficient for understanding the reference to the objects of any abstract mathematical structure.

Our reply is that the understanding of the general notion of reference, "however it works", rests necessarily on the presupposition that in some way (though it doesn't matter which) it must work. Thereof the question: Is there any way of referring to an arbitrary real number?

Shapiro is an upholder of the so-called *realism ante rem*, according to which mathematical objects are conceived of as positions in abstract structures, whose existence is prior to their possible specific instances. But, aside from the difficulty of a non-metaphorical understanding of what such positions are, if they are to be treated as objects of quantification, one cannot avoid the question: what does it mean to single out an arbitrary position?

1.7 Arbitrary Reference and Impredicativity

As we saw, Ramsey's criticism to Russell's thesis that an infinite class must be defined by a propositional function overlooks the problem of arbitrary reference, of which Russell was aware.

Ramsey's argument has been resumed by Gödel in his paper on Russell's mathematical logic. Gödel criticizes Russell's vicious circle principle, according to which no totality can contain members definable only in terms of the totality itself. Gödel observes that classical mathematics does not respect such a principle; and since classical mathematics can be reconstructed on the basis of the *Principia*, this work itself cannot respect that principle either "if 'definable' means 'definable within the system' and no methods of defining outside the system (or outside other systems of classical mathematics) are known except such as involve still more comprehensive totalities than those occurring in the systems". He adds:

> I would consider this rather as a proof that the vicious circle principle is false than that classical mathematics is false, and this is indeed plausible also on its own account. For, first of all one may, on good grounds, deny that reference to a totality necessarily implies reference to all single elements of it or, in other words, that "all" means the same as an infinite logical conjunction. (Gödel, 1984, p. 127)

Then Gödel observes that, even if "all" were intended as an infinite conjunction, the vicious circle principle would be tenable only within a constructive perspective:

> In this case [i.e. if the entities in question are constructed by us] there must clearly exist a definition (namely the description of the construction) which does not refer to a totality to which the object defined belongs, because the construction of a thing can certainly not be based on a totality of things to which the thing to be constructed itself belongs. If, however, it is a question of objects that exist independently of our constructions, there is nothing in the least absurd in the existence of totalities containing members, which can be described (i.e., uniquely characterized) only by reference to this totality. (Gödel, 1984, p. 127)

In a note he points out that "an object a is said to be described by a propositional function $\phi(x)$ if $\phi(x)$ is true for $x = a$ and no other object". (Gödel, 1984, p. 127)

Certainly for "definable" Russell doesn't mean "definable in the system". For, it is well known that those predicative propositional functions, whose existence is granted by the axiom of reducibility, fail, in general, to be definable in the object language of type theory: the range of the variables include functions indefinable in the object language. But, as we saw, the values of quantified variables must be capable of singular reference. It would, therefore, be circular to accept as range of quantification a universe with some members definable only by quantifying over the universe itself. Gödel's remarks don't take into account the sense of "definition" as ideal singular reference.

By contrast, **PAR** supplies a plausible justification of the vicious circle principle, provided that "definable" is interpreted as "capable of singular reference". For, as we saw, quantification over a universe $\mathcal{U}$ presupposes the possibility of arbitrary reference to any member of $\mathcal{U}$. So an act of arbitrary reference cannot involve in turn, on pain of vicious circularity, quantification over $\mathcal{U}$. Precisely, we can restate the vicious circle principle as follows:

(VCP) No universe of discourse can contain a member such that the agent can refer to it only by means of quantification over the universe itself.

Observe that, **VCP** is compatible with **CP** of second-order logic:

$$\textbf{(CP)}\ \exists X \forall y (Xy \leftrightarrow Ay).$$

For, in agreement with Russell's reducibility principle, one can imagine that the variable X ranges over a so rich universe of properties, satisfying **PAR**, that, some of them is *extensionally* equivalent to the propositional function Ay, even if the latter might involve a second-order quantification.

In other words, **VCP** is compatible with Russell's reducibility axiom. For, one can imagine that the agent has direct access to a certain universe of primitive properties (possibly non expressible in the formal language). If second order variables are restricted to such properties, **CP** can be accepted, without any circularity, as a richness assumption: it says intuitively that the universe of primitive properties is so rich that every extension of a second order propositional function is the extension of some primitive property.

References

Boyer, C. B., & Merzbach, U. C. (2011). *A history of mathematics*. London: John Wiley & Sons.

Breckenridge, W., & Magidor, O. (2012). Arbitrary reference. *Philosophical Studies*, *158*(3), 377–400.

Fine, K. (1985). *Reasoning with arbitrary objects*. OUP Press.

Frege, G. (1891). Function and concept. In P. Geach & M. Black (Eds.), *Translations from the philosophical writings of Gottlob Frege* (pp. 137–156). Oxford: Blackwell.

Frege, G. (1904). What is a function? In B. MacGuinness (Ed.), *Collected Papers in Mathematics, Logic and Philosophy* (pp. 285–292). Oxford: Blackwell.

Gödel, K. (1984). Russell's mathematical logic. In P. Benacerraf & H. Putnam (Eds.), *Philosophy of mathematics* (pp. 211–232). Cambridge: Cambridge University Press.

Hawthorne, J., & Manley, D. (2012). *The reference book*. Oxford: Oxford University Press.

Hintikka, J. (1996). *The principles of mathematics revisited*. Cambridge University Press.

Kripke, S. (1980). *Naming and necessity*. Oxford: Blackwell.

Martin, G. E. (2012). *The foundations of geometry and the non-Euclidean plane*. Berlin: Springer Science & Business Media.

Parsons, C. (1990). The structuralist view of mathematical objects. *Synthese*, *84*(3), 303–346.

Price, R. (1962). Arbitrary individuals and natural deduction. *Analysis*, *22*(4), 94–96.

Ramsey, F. (1925). The foundations of mathematics. *Proceedings of the London Mathematical Society*, *25*, 338–394.

Russell, B. (1908). Mathematical logic as based on the theory of types. In J. van Heijenoort (Ed.), *From Frege to Gödel* (pp. 150–182). Cambridge MA: Harvard University Press.

Shapiro, S. (1997). *Philosophy of mathematics*. Oxford: Oxford University Press.

Chapter 2
On Plural Arbitrary Reference

Abstract This chapter introduces a new approach to plural quantification through the concept of *plural arbitrary reference*. It highlights the implicit presupposition in mathematical reasoning that any individual in the universe of discourse can be referred to. By introducing a team of ideal agents capable of direct access to any individual, plural arbitrary reference is achieved through simultaneous acts of choice by each agent. This idealized notion of reference provides a basis for understanding plural quantification without second-order entities, maintaining the ontological innocence of second-order logic by distinguishing between acts and entities.

2.1 Introduction

In this chapter we propose a new approach to plural quantification via a highly idealized notion of *plural arbitrary reference*.

Our approach will start from the observation that the possibility, in principle, of referring to any individual of the universe of discourse of a mathematical theory is essentially, although implicitly, presupposed in mathematical reasoning. In order to make such presupposition explicit and to clarify the sense of the locution "in principle," we are going to introduce a *team of ideal agents*, which are supposed to have direct access to any individual. Plural arbitrary reference to certain individuals will be realized through an act of plural choice, i.e. an act of choosing an individual performed simultaneously by each agent. By means of this device, we will obtain an idealized version of Kripke's notion of direct reference, more basic than any sort of reference by description. Plural quantification will be explained in terms of such notion of reference: *plural arbitrary reference*.

We think that our framework provides a new perspective for understanding the nature of *plural arbitrary reference* and *plural quantification*.

Pluralities of individuals are not second-order entities: speaking of a *plurality* of individuals is nothing but a linguistic device for referring to the individuals selected

M. Carrara, E. Martino, *Arbitrary Reference in Logic and Mathematics*, Synthese Library 490, https://doi.org/10.1007/978-3-031-66452-6_2

by an act of choice. Moreover, our team of agents will provide an intuitive means for simulating relations, so that *full* second-order logic can be reconstructed without any further device. We think that a careful distinction between acts and entities can vindicate the ontological innocence of second-order logic.

2.2 Boolos on Second Order Monadic Logic

Boolos in (1984, 1985) has proposed an interesting reinterpretation of second-order monadic logic in terms of *plural quantification*. In Boolos' perspective, second-order monadic logic is ontologically innocent: second-order variables do not range over sets of individuals but over individuals *plurally*.

Boolos' basic idea consists of interpreting the atomic formulas of the form Xy, as "y is one of the Xs," and the existential formulas of form $\exists X \ldots$ as "There are some individuals Xs such that… ." The universal quantifier $\forall X$ is expressible in terms of the existential one in the usual way. For instance, consider the second-order definition of an ancestor:

a is an ancestor of $b =_{df.} \neg\exists X(Xb \wedge \forall x\forall y(Xx \wedge Pyx \rightarrow Xy) \wedge \neg Xa)$,

where Pyx is the relation "y is a parent of x." (For simplicity, b himself is included among his ancestors.) In Boolos' interpretation, this is to be read:

"There are no individuals Xs such that (1) b is one of the Xs, (2) every parent of one of the Xs is one of the Xs, (3) a is not one of the Xs."

Boolos gives no explanation of how to refer to an arbitrary plurality of individuals. He treats directly *plural existential quantification* taking as primitive the locution "there are some objects such that…" in natural language. But the meaning of this locution is somewhat ambiguous, strictly depending on the context of discourse. In some contexts, its meaning is the same as the first-order quantification "there is at least an object such that…". And when such locution is not reducible to a first-order quantification, as in the famous Geach-Kaplan's proposition

"Some critics admire only one another",

it may seem a sloppy way of referring to some *class* of individuals. This proposition, not formalizable in first-order language, seems, according to Resnik, hardly interpretable without resorting to classes. How could we understand "one of them" without referring to a certain class and agreeing that the referent of "one" belongs to it? (Resnik, 1988). Besides, the natural language does not suggest a direct interpretation of the universal quantifier: "for each individuals…" is ungrammatical and "for all individuals" is indistinguishable from the first-order quantification "for each individual."[1] Boolos disregards the problem and defines universal quantification in terms of existential and negation in the usual way.

Truly, Boolos provides also a formal semantics for his language in *Nominalistic Platonism* (Boolos, 1985). He gives a semantic for second order logic restating the Tarskian truth definition by modifying the notion of *assignment* as follows. Given a domain D of individuals, he defines as an assignment any binary relation R between variables and individuals that correlates a unique individual with every first order variable, while it is subject to no constraint for second order variables. So R may correlate a second order variable with no, one or (possibly infinitely) many individuals. The *satisfiability* relation is inductively defined as usual, with the following clauses for atomic formulas and second order existential quantification:

(i) R satisfies the atomic formula Fx *iff* the correlate of x is one of the correlates of F;
(ii) R satisfies $\exists FA$ *iff* there is a relation R', differing from R at most for the correlates of F, such that R' satisfies A. (The universal quantifier is defined in terms of the existential one).

Truth is then defined, as usual, in terms of *satisfaction*. So the *set* of the correlates of F is not involved in the definition of truth. This makes the notion of plural quantification precise and shows how it yields an alternative semantics for second order logic. This semantics turns out to be equivalent to the usual one, according to which the values of second order variables are all sets of individuals. And since the notion of *value of a variable* can be made precise only by the definition of *assignment*, in Boolos' perspective the proposed reformulation shows that, using Quine's slogan that "to be is to be the value of a variable", there is no commitment in second order logic to any entities but individuals.

Of course, the new definition of assignment presupposes the notion of set of individuals if relations are understood set-theoretically. But a relation can in turn be understood in terms of *plural arbitrary reference* to certain ordered pairs (taking for granted the notion of ordered pair).

However, the use of plural reference in the meta-language begs the crucial question, whether plural reference involves surreptitiously the notion of set. Boolos' view, although very attractive, has met the criticism of several philosophers of mathematics. The main criticism rests on the suspicion that speaking of pluralities of individuals is just a rough manner of speaking of sets. Take, for example, Parsons' criticism to Boolos (Parsons, 1990). Although he acknowledges to Boolos the merit of throwing new light on the old notion of manifold, he argues that his interpretation of second order logic is not ontologically noncommittal:

[1] Lewis has suggested an interpretation of the universal plural quantifier as "whenever there are some things, then..." (Lewis, 1993, p. 11). But, we have some doubt that such locution is appropriate. "Whenever there are some numbers, then..." sounds somewhat strange and seems to suggest that it might happen sometimes that there are no numbers.

> The great interest of his reading, [. . .], is that he breathes new life into the older conception of pluralities or multiplicities. As a source of second order logical forms, the plural and plural quantification are rightly distinguished from what was so much emphasized by Frege, predication and, more generally, expressions with argument places. In particular, if it is the idea of generalization of predicate places that we appeal to in making sense of second-order logic, then the most natural interpretations will be relative substitutional or by semantic ascent, and these will not license impredicative comprehension, and it is hard to see how that will be justified. But if one views examples such as Boolos's as involving 'pluralities', they are more like sets as understood in set theory in that no definition by a predicate is indicated, so that one need not expect them to be definable at all. Thus no obstacle to the acceptance of impredicative comprehension is removed.
> An advocate of Boolos' interpretation in an eliminative structuralist setting could grant my claims about ontological commitment, but then take a position analogous to the Fregean: second-order variables indeed have pluralities as their values, but these are not objects. It does not seem to me to have the same intuitive force as Frege's position, since there is no analogue to the regress argument that can be made if one views the reference of a predicate as an object. There will still be, just as with Frege's concepts, the irresistible temptation to talk of pluralities as if they were objects, as we have already noted above. The only gain this interpretation offers over the Fregean is a more convincing motivation of impredicativity. (Parsons, 1990, p. 328)

Quine's old claim that second-order logic is *set theory in disguise* (metaphorically "a wolf in sheep clothing") does not seem to have lost all its advocates. We propose to supersede the above criticism with a new approach to plural quantification. We will argue that plural quantification in logic (and mathematics) can be better understood within the frame of a highly idealized notion of *plural reference*.

2.3 Plural Reference: Some Background

The idea of explaining the notion of a set by means of plural reference can be found, in nuce, in Black's famous paper *The elusiveness of sets* (Black, 1971). Black starts from the consideration that, although the basic notions of set theory are nowadays very familiar to all mathematicians, the very nature of a set, conceived of as a well-determined entity built up from its members, is quite mysterious:

> Beginners [of set theory] are taught that a set having three members is a single thing, wholly constituted by its members but distinct from them. After this, the theological doctrine of the Trinity as "three in one" should be child's play. (Black, 1971, p. 616)

Black criticizes at length Cantor's well-known definition of set: By a "set" we understand every comprehension *S* of determinate well-distinguished objects of our intuition or thinking into a whole, as well as all other attempts to explain what a set, as an abstract entity, is. He regards such attempts as a misleading mystification and recommends avoiding any question about the nature of sets. What is important, according to Black, is rather to look at the use of *plural reference* in natural language as the appropriate starting point for learning to master the sophisticated language of mathematical set theory:

> The notion of "plural" or simultaneous reference to several things at once is really not at all mysterious. Just as I can point to a single thing, I can point to two things at once, using two hands, if necessary; pointing to two things at once need be no more perplexing than touching two things at once. Of course it would be a mistake to think that the rules for "multiple pointing" follow automatically from the rules for pointing proper; but the requisite conventions are almost too obvious to need specification. The rules for "plural reference" are no harder to elaborate. (Black, 1971, p. 629)

Black's conclusion can be satisfactory for a working mathematician who is only interested in the correct use of a set-theoretical talk. But it is unsatisfactory for a philosopher interested in understanding how to think of the objects, if any, that set theory seems to speak of. We think that, in a philosophical perspective, "the requisite conventions for multiple pointing" are far from being obvious and need an adequate specification, which we want to propose in the sequel.

Another source of the idea of a set of natural numbers as generated by a plurality of choices is Bernays in "On Platonism in Mathematics":

> The value of platonistically inspired mathematical conceptions is that they furnish models of abstract imagination. These stand out by their simplicity and logical strength. They form representations which extrapolate from certain regions of experience and intuition.... [Platonism] abstracts from the possibility of giving definitions of sets, sequences, and functions. These notions are used in a "quasi-combinatorial" sense, by which I mean: in the sense of an analogy of the infinite to the finite....[e.g] one views a set of integers as the result of infinitely many independent acts deciding for each number whether it should be included or excluded. We add to this the idea of the totality of these sets. (Bernays, 1983, p. 259)

Of course, one may strengthen the idealization and imagine that the agent is able to perform infinitely many actions simultaneously. In this perspective, Velleman (Velleman, 1993) proposed his "liberalized constructivism".

He observes that, according to intuitionism, classical logic can be used when reasoning about finite domains, but quantification over infinite domains requires the use of intuitionistic logic. In contrast, liberal constructivism proposes that the line drawn by intuitionists between finite and infinite be drawn instead between countable and uncountable. Velleman considers Dummett's arguments for rejecting classical logic both for natural numbers and for real numbers. The idea is that classical logic is appropriate for quantification over well determined domains, while natural and real numbers are indeterminate, since the notion of natural number is vague and real numbers are indefinitely extensible. In spite of the well determined generative process of natural numbers, the vagueness of the notion of natural number would be suggested, according to Dummett, by Goedel's incompleteness theorem:

> If it is acknowledged that the definiteness of the extension of "natural number" does not imply definiteness for the notion of a ground for asserting something to be true of all natural numbers – for the meaning of quantifiers whose variables range over the natural numbers – there is no longer any reason for resisting the idea that in this respect the notion of natural number is an inherently vague one. (Dummett, 1963, p. 197)

But, as Velleman observes, the lack of a ground for asserting any arithmetical sentence by no means can lead to doubt the intelligibility of classical quantification over natural numbers. What is to be determined by the meaning of quantification is not the ground for asserting any arithmetical sentence but the classical truth value of any arithmetical sentence. And Velleman maintains that such truth value is determined according to a liberal conception of computation. For, Intuitionism idealizes the notion of computation by imagining an ideal agent, free of spatio-temporal limitations, who is able to execute any finite, however long, computation. On the other hand, the intuitionistic idealized agent cannot execute in a finite time any computation requiring an infinite number of steps. Following an idea of Russell, Velleman sees such limitation as merely "medical" and liberalizes the ideal agent by imagining him free of that limitation. Accordingly, the ideal agent is able to compute the truth value of any first order arithmetical sentence. Now, we want to observe that, once accepted the possibility of completing an infinite computation, liberal constructivism is committed to the actual infinite.

In contrast, as we will see, according to our idealization, any computation, however long, is to be finite. In this vein, we refrain from assuming that a single ideal agent be able to perform a plural choice, consisting of infinitely many simultaneous choices. For the purpose, it seems more accessible to human imagination a team of infinitely many agents. Our agents are imagined as concrete humans; each of them is able to choose any individual (or finitely many individuals) ad libitum. Since, according to mathematical structuralism, the nature of individuals is indetermined, these can be imagined as concrete entities. Besides, for our purposes, they can be identified with the agents themselves. Concreteness guarantees the possibility of direct reference by ostension. Each agent has direct access to any agent. The working mathematician can imagine to be the leader of the team: at any moment he can charge any agent of choosing an individual ad libitum, as well as he can charge the whole team to a specific act of choice. Besides he imagines to have no knowledge of the outcomes of the choices at issue. Nevertheless he can refer to the chosen individuals merely through the supposed knowledge of the acts of choosing.

Another important insight to obtain an adequate specification of plural referring comes from Kitcher, who has elaborated a conception of set theory in terms of *ideal actions* (see Kitcher, 1984). Kitcher has clearly realized the importance of an *idealized* notion of *act* vs. that of *object*. He has proposed a reconstruction of Mill's arithmetic and of *ZF* set theory in terms of the acts of *collecting* of an *ideal agent*, free of the empirical limitations of real human beings:

> One central idea of my proposal is to replace the notions of abstract mathematical objects, notions like that of a *collection*, with the notion of a kind of mathematical activity, *collecting*.... In its most rudimentary form, collecting is tied to physical manipulation of objects.... We learn how to collect by engaging in this type of activity. However, our collecting does not stop here. Later we can collect the objects in thought without moving them about. We become accustomed to collecting objects by running through a list of their names, or by producing predicates which apply to them. Naively, we may assume that the production of any predicate serves to collect the objects to which it applies. (This naive assumption is implicit in nineteenth-century analysis, and it was made explicit by Cantor.) Thus, our collecting becomes highly abstract. We may even achieve a hierarchy of

> collectings by introducing symbols to represent our former collective activity and repeating collective operations by manipulating this symbols.... As I construe it, the notation '{...}' obtains its initial significance by representing first-level collecting of objects, and iteration of this notation is itself a form of collective activity. (Kitcher, 1984, p. 110)

So Kitcher's idealized *collecting* seems to be the product of an extrapolation from several ways of collecting in our real life. Unfortunately, he deliberately doesn't explain what the ideal act of collecting consists in:

> Obviously, the ideal subject is an idealization of ourselves, but I explicitly reject the epistemological view that we can know *a priori* the ways in which the idealization should be made. (Kitcher, 1984, p. 111)

We think, however, that the rejection to pinpoint the idealization in question precludes the possibility of supplying a well-determined conception of the mathematical infinite and of justifying the adoption of classical logic.

How can the ideal agent collect infinitely many objects? Is this suggested by our manipulations of concrete objects? This familiar activity may suggest that, if we were free of spatio-temporal limitation, we could indefinitely collect more and more objects. But could an act of collecting consist of an indefinitely proceeding process?

To admit an infinite process of collecting, step by step, new arbitrarily chosen objects would make undetermined whether any given object will be collected or not in the course of the process. But Kitcher assumes, in accordance with classical logic, that it is well determined, for any *collective operation*, which objects are collected. Nor can an infinite collection be conceived of as the result of singling out infinitely many objects by means of a common property: collecting should be a primitive notion not grounded on the highly problematic notion of property. Kitcher is aware of this fact and holds that a collective operation is performed independently of any property we use for describing the collected objects. But then, the mere assumption that the ideal agent is free of empirical limitations is far from giving any account of how the infinitely many actions of the ideal agent are to be imagined. Concerning this point, Brown observes:

> Not surprisingly, the ideal agent is faster than a speeding bullet, stronger than a mighty locomotive, and able to leap tall buildings in a single bound. This much idealization is rather harmless and Kitcher is certainly entitled to help himself to it. (Superman, after all, is still a finite being). But when it comes to infinite operations, we must surely object. This cannot be passed off as merely overcoming an "accidental limitation" that the rest of us humans have. A platonic realm is not half so mysterious or implausible. (Brown, 2003, pp. 11–12)

Just in order to supersede this difficulty, we will introduce, in Sect. 2.4, a *team of infinitely many agents*.

Kitcher regards his philosophy of mathematics as a *philosophical naturalism*, since it starts from considering certain usual actions of real life. S. Hoffman (Hoffman, 2004) observes that, since the ideal agent does not exist, Kitcher's philosophy would be better understood as a form of *fictionalism*. We think, however, that a fictional aspect, together with the naturalistic one, is already implicit in Kitcher's perspective, since any idealization is fictional in the obvious sense that idealized entities do not exist and idealized actions are not performable by real human beings.[2]

But if one wants to go fictionalist, why should one not prefer a fictionalism about mathematical objects rather than about operations? One might argue that the former shares with Platonism the advantage of taking mathematical language at face value, while the latter seems to be somewhat bizarre. Hoffman defends the ideal-agent-fictionalism as follows:

> The ideal-agent formulation takes mathematics to be about operations rather than abstract mathematical objects. If mathematics is really about abstract objects then it is entirely false, since there are none of these at all. If mathematics is a set of stories about operations, however, then a portion of it that is about operations that actually exist really is true.... Operations are kind of things that really exist; numbers are not.... Mathematics is not only a tool, it is also a theory of what mathematicians *do*. (Hoffman, 2004, p. 14)

This kind of argument is not convincing. We don't think that the non existence of mathematical objects outside of our imagination makes mathematics false. If mathematical talk is about fictional objects, then mathematical truth is to be understood as concerning only the fictitious world of such objects. The truth of the proposition

"there are infinitely many prime numbers"

by no means implies that prime numbers are non-fictional objects. In general, no mathematical proposition entails that the existence of mathematical objects is independent of our game of make-believe. As to the concrete mathematical activity, one could object that it does not consist of singling out and ordering concrete objects. Mathematicians do not perform such actions, they rather reason on notions grasped by reflecting on such kind of actions. Mill's and Kitcher's interpretations of the mathematical talk as about actions is certainly somewhat artificial, since it is plain that mathematical talk, taken at face value, is about objects.

However, we want to argue in the next section that mathematical reasoning about objects implicitly presupposes a highly idealized possibility of referring arbitrarily to them, and that the introduction of ideal agents can serve the purpose of making such an ideal reference explicit. In this sense, we want to embrace a sort of fictionalism about objects that involves a fictionalism about ideal agents.

---

[2] For an introduction to *fictionalism* in the philosophy of mathematics see (Balaguer, 1998, 2009; Yablo, 2002; Thomas, 2000, 2002).

2.4 Plural Arbitrary Reference

In Chap. 1 we have introduced the notion of *arbitrary reference*.

Question: How can it be applied to the semantics of second-order logic?

Since, according to the classical semantics, second-order variables are intended to range over all classes of individuals, the problem in terms of our frame of arbitrary reference is reformulated in the following terms:

Question: How can the ideal agent have direct access to any class of individuals?

Since all we know about an arbitrary class is that it is an entity determined by its members (or by absence of members, in the case of the empty set), it seems that one can have access to a class only through its members. And since a class may be infinite, it seems that a single agent is unable to perform a simultaneous choice of all members of any class, unless he is conceived of as a horrible monster with infinitely many fingers. We don't like that!

A choice of infinitely many individuals, *one at a time*, would face the hard problem of dealing with undetermined classes (as it happens in intuitionistic mathematics). Black himself describes a "plural pointing" as a "*simultaneous* reference to several things at once." Kitcher seems to disregard this difficulty (see Kitcher, 1984, chap. 6). He assumes that his ideal agent is able to collect any infinity of objects. Moreover, as already observed, he outlines a reconstruction of set theory by assuming that the ideal agent is able to iterate his collecting operations over and over reproducing the whole transfinite hierarchy of types required by the iterative conception of set.

What does it mean to collect some actions of collecting? Kitcher's answer is that, once certain collecting acts have been performed, one can introduce some symbols for representing them and perform collecting acts of such symbols. So one should understand the collecting of acts as a collecting of representative of acts. This explanation is certainly useful for didactic purposes: it can help a student to understand what is a set of sets. But it is inadequate from a philosophical point of view. Moreover, it is misleading within Kitcher's framework. In fact, that explanation suggests to the student that one can collect sets because sets are in turn objects. That is true according to the usual conception of set, but Kitcher's collecting acts fail to be objects. Kitcher himself maintains that an operation exists only insofar as it is performed. So the expedient of introducing symbols as representative of acts cannot give any account of how acts can be in turn collected. In our framework there will be no room for collecting acts.

2.5 A Team of Agents Making Choices for a Semantics of Acts of Choices (SAC). How Plural Arbitrary Reference Works

We propose to overcome the discussed difficulties by extending the idealization of a single agent introducing an *infinite* team of agents, consisting of a number of agents equal to the number of individuals. Yet, for our purposes, we can take as individuals the agents themselves. This idea implements our conception of *plural arbitrary reference*. Our agents are supposed to be able to perform the following actions:

1. Singular selecting choice (*s.s.c.*): one of the agents chooses an individual ad libitum;
2. Plural selecting choice (*p.s.c.*): it is performed by all agents *simultaneously*: each agent chooses an individual ad libitum (independently one of the others) or refrains from choosing;
3. Plural relating choice (*p.r.c.*) of degree $n \geqslant 2$: it is performed by all agents simultaneously: each agent chooses n (not necessarily distinct) individuals in a certain order or refrains from choosing. (Abstention from choosing serves the purpose of introducing empty pluralities and relations).

Imagine that the team is guided by a real mathematician, who can order at will the execution of one of the foregoing actions. By means of such actions, he can refer to a single individual or to a plurality of individuals or to a *plural relation* among individuals, without submitting to abstract entities the job of collecting and correlating individuals.

A locution as:

"Let X be an arbitrary plurality of individuals"

is to be rephrased as:

"Suppose we have ordered a *p.s.c.* and call X the chosen individuals."

So the locution in question is to be understood as a description of an act of reference performed by us, relative to a certain plural choice. Similarly, a universal quantification:

"for every plurality X"

is to be read as:

"however a *p.s.c.* of certain individuals X is performed";

an existential quantification:

"there is a plurality such that "

is to be read as:

"it is possible that such a *p.s.c.* of certain Xs be performed that."

Formally we use a full second-order language L with identity, with first-order variables x, y, z and second-order variables X^n, Y^n, Z^n (of any degree $n \geqslant 1$). We omit the superscripts for variables of degree 1.

Let us explain the *semantics of acts of choice* (**SAC**).

An *assignment* to a formula A for every free variable v (of any sort) in A, is an appropriate act of choice, i.e., a $s.s.c.$ for every first-order variable, a $p.s.c.$ for every second-order variable of degree 1, a $p.r.c.$ of degree n for every variable of degree $n \geqslant 2$. With reference to an assignment to a formula A, if v is a free variable of A of any sort, we indicate by v^* the relative act of choice.

We define inductively the truth of a formula relative to an assignment:

(i) $x = y$ is true if x^* and y^* choose the same individual;
(ii) Xy is true if the individual chosen by y^* is one of the individuals chosen by X^*;
(iii) $X^n y_1 \ldots y_n$ is true if the individuals chosen respectively by $y_1^* \ldots y_n^*$ are chosen in the order by X^{n^*};
(iv) usual clauses for the propositional connectives;
(v) $\forall v B$ is true if, however the assignment may be extended to B by an appropriate act of choice v^* for v, B turns out to be true;
(vi) $\exists v B$ is true if it is performable an act of choice v^* for v such that B turns out to be true.

Now, it is clear that a $p.s.c.$ does not create any entity that collects the chosen individuals. Speaking of pluralities as if they were genuine entities is a mere *façon de parler*, paraphrasable in terms of plural choices. Thus, *plural arbitrary reference* does not involve the notion of class. On the contrary, singular reference to a class presupposes plural reference to its members. In this sense, Boolos' claim that second-order logic, interpreted in terms of plural quantification, does not involve second-order entities is vindicated.

Boolos' approach to plural quantification provides an interpretation only of monadic second-order logic, because his appeal to natural language does not help to simulate relations. Lewis, in turn, extends the interpretation to full second-order logic, by using mereological sums of individuals for codifying ordered pairs of individuals. An advantage of our approach—as we are going to argue in the second part of this book—is that we do not need mereology in order to recover full second-order logic, because our team of agents provides obvious means to simulate relations of any degree. A $p.r.c.$ of any degree is not more problematic than a $p.s.c.$, since choosing n individuals in a certain order is a finite operation not more problematic than choosing a single individual. In this way, one can better appreciate the power of our device of plural choices. This will be analysed in the second part of the book.

Our interpretation of second-order logic cannot be extended to third-order logic. In fact, since pluralities are not objects and have no actual existence, there is no room for referring, through an act of choice, to an arbitrary plurality of pluralities of individuals. One can try, however, to take seriously Kitcher's suggestion of collecting pluralities by collecting certain symbols representative of them. More precisely, for any binary *p.r.c.*, one can take each individual as a representative of all individuals correlated with it. Then one can wonder if a *p.r.c.* is possible that introduces a representative of every possible plurality. But the answer is negative, according to the following plural version of Cantor's theorem:

(Cantor) $\neg\exists X^2\forall Y\exists x\forall y(Yy \leftrightarrow X^2xy)$

Proof Suppose, by way of contradiction, that X^2 is such that $\forall Y\exists x\forall y(Yy \leftrightarrow X^2xy)$. By (**CP**) there is a plurality Z such that $\forall x(Zx \leftrightarrow \neg X^2xx)$. Let z such that $\forall x(Zx \leftrightarrow X^2zx)$. We get the contradiction $(Zz \leftrightarrow \neg Zz)$. ■

The theorem shows that, although, through a binary *p.r.c.*, the leader can refer to an infinity of pluralities, he cannot refer to all *possible* pluralities. In other words, the potential existence of pluralities cannot be completely actualized: no act of choice can give an actual simultaneous existence to all possible pluralities.

2.6 A Fictional Constructivism About Acts of Choice

The fictional Platonist flavour of our theory consists in the fact that individuals (i.e. our agents) are treated as if they actually existed. That is essential in order to make acts of plural *simultaneous* choices performable. In this respect, our perspective could be labelled as a "fictional Platonism about agents". In contrast, acts of choice, unlike agents, are to be understood in a mere *potential* way: there is no realm of possible acts; an act exists only insofar as it is performed. In this respect, our perspective could be labelled as a "fictional constructivism about acts of choice."

On this point, it is worth clarifying how to understand the notion of *possibility* involved when speaking of *possible* acts of choice. As already observed, we don't assume any ontology of possible acts. If possible acts were understood as entities of a realm of *possibilia*, then the problem of arbitrary reference to the objects of the universe of discourse would be simply reduced to the even harder problem of arbitrary reference to *possibilia*. The force, if any, of our approach rests essentially on the view that acts of choice are no entities at all, neither actual nor possible. The possibility in point is merely *combinatorial* and *non-epistemic*: it is determined by the rule of the play that establishes what every agent is allowed to do when performing an act of choice. This notion of possibility is perfectly compatible with the use of classical logic. For, one can recognize, by induction on the complexity of a formula, that the truth-value of a formula, relative to an assignment, is well-determined by the truth clauses. As an example, let us consider the case of an existential formula $\exists XB$. An instance of clause (vi) says that:

$\exists XB$ is true if it is performable a *p.s.c.* X^* for X such that B turns out to be true.

By the induction hypothesis, however a *p.s.c.* for x may be performed, it determines a truth value of B. The combinatorial possibilities concerning the performance of a *p.s.c.* are determined by the play rule governing plural choices, according to which every agent is allowed to choose an individual ad libitum. Hence it is well determined if the possibility is left that a *p.s.c.* be performed in such a way that B turns out to be true. Thus, $\exists X B$ has a well-determined truth-value. Besides, our notion of combinatorial possibility justifies immediately, without any circularity, the comprehension principle of second-order logic:

(CP) $\exists X^n \forall y_1 \ldots y_n (X^n y_1 \ldots y_n \leftrightarrow A)$.

For, in virtue of the arbitrarity of choices allowed by the choice rule, nothing can prevent the possibility of a *p.r.c.* (or a *p.s.c.*) such that the chosen n-tuples are just the ones satisfying A. Possible occurrences in A of second-order quantifications cannot produce any circularity, since acts of choice, unlike properties, are all independent one of the other.

As we saw, our notion of truth is *absolutely non-epistemic*. Instead, on this respect, Kitcher's perspective seems to be rather ambiguous. He criticizes the intuitionistic view of mathematics as based on a certain confusion between the ontological and the epistemic aspect (Kitcher, 1984, p. 143). Kitcher claims that his proposal wants to develop the constructive ontological thesis, while repudiating the epistemic one. Intuitionism aims to avoid the classical notion of truth by grounding the meaning of mathematical propositions on the notion of assertability. However, Kitcher observes, assertability involves the notion of verification, which, in turn, involves the notion of truth. We entirely agree with this objection. But then, according to this objection, Kitcher's notion of truth would be expected to be non-epistemic. However, comparing his ideal agent with Brouwer's creative subject, Kitcher justifies his use of classical logic by removing the intuitionistic limitations to the knowledge of the creative subject and adopting a more liberal idealization about the epistemic power of the ideal agent (Kitcher, 1984, pp. 144–145). This move suggests that his conception of mathematical truth is still, after all, of an epistemic nature. As for the Brouwerian intuitionist, a proposition is true only if it is verifiable by the ideal agent.

2.7 On the Ideal Agent, Again

We think that human imagination has an essential role both in the classical and in the constructive conceptions of mathematics. The difference between Platonism and constructivism amounts to a difference of idealizations, i.e. to different ways of imagining the mathematical universe. While this is thought of by a platonism as an autonomous realm preexistent to the mathematical activity, it is imagined by an intuitionist as generated along space and time by idealized human actions. The introduction of an ideal agent has the purpose of making perspicuous certain actions that the working mathematician recurrently imagines to be able to do "in principle". In particular, we are interested in the action of arbitrary reference.

Our ideal agent is a fictional character thought of as a human being, free of space-time limitations, who can choose ad libitum any individual of the universe of discourse. A very simple way of referring to a concrete object in everyday life is by ostension. This suggests to imagine that the agent can choose any individual by ostension.

One may wonder if mathematical entities are capable of been indicated by ostension. One may object that, dealing with abstract entities, as infinite sets, direct reference by ostension is impossible. That is certainly right for abstract mathematical objects, as conceived of by Platonism. However, the structuralist conception of mathematics points out that a mathematical theory does not determine the very nature of the objects it speaks of. That opens the way to imagine that, given any mathematical theory, we can reason about a model, whose domain, however infinite, consists of concrete objects. For instance, as a model of natural numbers, we can take Hilbert's finite strings of strokes. Each of these, however long, is imagined as directly accessible to the agent. When dealing with second order logic, we have introduced a team of infinitely many agents as a device for choosing the members of any set of natural numbers, avoiding the reification of sets.

It is worth noting that an "arbitrary object", as here understood, is not to be confused with a "generic object". The latter does not belong to the domain D at issue, but it is an abstract entity enjoying the features shared by all members of D. It is a very puzzling entity. As observed by Locke, a generic triangle "must be neither oblique, nor rectangle, nor scalenon; but all and none of these at once".[3] By contrast, our "arbitrary object" is to be understood as a well determined member of D. The adjective "arbitrary" does not attribute any quality to the object in question, but qualifies the act of referring. In our perspective, to say that *a* is an arbitrary object means that we have charged our ideal agent to choose at a certain moment a member of D ad libitum and that we call "*a*" such an object, unknown to us. Since all that we now about *a* is that it is a member of D, it is clear that any property that we can prove for *a* must hold for all members of D. This justifies the introduction rule of universal quantification. One may wonder if the role of an arbitrary object could be played by a generic object. We think not. For, since the contradictory properties attributed to a generic object are not shared by any member of D, universal quantification seems to be unjustified.

Our agent, unlike the Brouwerian creative subject, is not expected to have any mathematical competence: he has no job other than ordering acts of choice to his team and referring to the chosen individuals, in particular to make assignments. Once an assignment to a formula has been made, the truth-value of the formula is well-determined by the clauses formulated above. Such value may not be known by any real or ideal subject. In this sense, our notion of truth is classical and presupposed by the notion of proof. Mathematical proofs are made by the working mathematician, who tries to discover truths by reasoning about individuals; the help of the ideal agents in the proof consists in the fact that the working mathematician, in the course of his reasoning, can exploit, though implicitly, the ability of the ideal

---

[3] The passage is in J. Locke, *An Essay Concerning Human Understanding* Oxford: Clarendon Press, 1975, book IV, chapter VII, §9.

leader of referring, singularly and plurally, to individuals. So our fictionalism is in agreement with the requirement that mathematical talk be taken at face value. In our framework, mathematical talk is about individuals (and not about actions, as in Kitcher's and Mill's interpretations) and uses ideal actions as a means of referring to them.

2.8 On Our Fictionalism

Mathematical fictionalism has been criticized by several authors. In his *Fictionalism, Theft, and the Story of Mathematics* (Balaguer, 2009, p. 152) paraphrases Burgess' criticism as follows:

> Fictionalists face a dilemma: they have to endorse either hermeneutic fictionalism or revolutionary fictionalism, but neither is plausible. We can define hermeneutic fictionalism as the view that mathematicians (and perhaps ordinary folk) intend their mathematical talk to be taken as a form of fiction; more specifically, the view here is that, according to ordinary mathematical intentions, singular terms like "4" are not supposed to refer, and sentences like "4 is even" are not supposed to be literally true. But hermeneutic fictionalism is wildly implausible and wholly unmotivated; as an empirical hypothesis about what mathematicians intend, there is no good evidence for it, and it seems obviously false. Revolutionary fictionalism, on the other hand, is the view that (a) mathematicians do not intend their utterances to be taken as fiction, or as non-literal in any other way; and so (b) we should interpret mathematicians as really asserting what their sentences say, i.e. as making assertions that are about (or that purport to be about) mathematical objects; but (c) since there are no such things as mathematical objects, the assertions of mathematicians are simply mistaken. But revolutionary fictionalism is pretty hard to swallow; for given the track records of philosophers and mathematicians, it would be comically immodest for philosophers to presume that they had discovered a problem with mathematics. (Burgess, 2004, p. 20)

As Balaguer observes, the argument against revolutionary fictionalism rests on the following empirical hypothesis:

Hermeneutic realism: Typical mathematicians (and perhaps ordinary folk) intend their mathematical utterances and theories to provide true descriptions of actually existing objective entities.

Under this hypothesis, Burgess maintains that it is "comically immodest" for a philosopher to dictate to mathematicians what they should say about mathematical questions. But, Balaguer continues, there is no evidence of such hypothesis. Besides, even if it holds, there is nothing of immodest in the disagreement of a philosopher with a mathematician since the point at issue is not mathematical but philosophical. No mathematical result is rejected by a fictionalist.

Certainly, a working mathematician, insofar as he argues about mathematical theories, reasons as if mathematical entities were existent and mathematical propositions were true or false. But that has nothing to do with her personal opinions about the existence of mathematical entities in the real world. About that she may even not have any opinion at all. Similarly, concerning hermeneutic fictionalism, any mathematician can agree that the truth values of mathematical sentences are determined by the history of mathematics. The question whether a mathematical theory tells the history of real entities or of merely fictional entities is irrelevant for mathematical works.

Now, we want to consider the fictional aspect of our theory of ideal agents. While the question of the real existence of numbers or, in general, of mathematical objects is debatable, it is plain that the existence of our ideal agents is merely fictional. Take, for the moment, the case of the single agent Kurt. This was introduced for explaining the meaning of recurring locutions, in mathematical theories, having the form:

(*) "Let a be an arbitrary object of the domain at issue".

Of course no revolutionary fictionalism (in Burgess'sense) is involved in such device: since no explanation of sentence (*) is provided by the working mathematician, no conflict can arise from our proposal.

Is our proposal a form of hermeneutic fictionalism? No, at least in Burgess' sense, since we don't maintain that any appeal to an ideal agent occurs in the intentions of the working mathematician. Perhaps the latter thinks that no explanation of locution (*) is needed, because the way of picking the arbitrary object a is mathematically irrelevant.

However, as we saw, the assumption of **PAR**, the principle of arbitrary reference, is exploited in logical deduction, so that the mathematician, who recognizes the correctness of natural deduction, must agree that, in some way, arbitrary reference is performable. So our appeal to the ideal agent has the purpose of proposing a way of understanding the correctness of a form of reasoning adopted by mathematicians. We think, therefore, that our fiction is a form of hermeneutic fictionalism, in the traditional sense of hermeneutics, understood as the art of interpretation. In fact, all that we do is to propose an interpretation of the locution (*), in particular of the term "arbitrary", as a clarification of its mathematical use.

As to the team of infinitely many agents: it provides an interpretation of second order logic free of second order entities. Of course, we are far from holding that the appeal to a team of agents is in the intentions of working mathematicians, so that we don't face Burgess' criticism to *hermeneutic fictionalism*. Nevertheless, such appeal is in full agreement with classical second order formal logic, as well as with Boolos' thesis of the ontological innocence of second order logic. We are not liable to Burgess' criticism against *revolutionary fictionalism*, since we don't claim that anything is wrong in the usual interpretation of second order logic.

What we have shown is only that, in order to endorse second order logic, the adoption of second order entities is overwhelming, so that, in particular, one can accept second order arithmetic within the mere conception of countable infinite.

We think, therefore, that even our appeal to an infinite team of agents may be seen as a form of *hermeneutic fictionalism*, insofar as it provides a new interpretation of second order logic of some relevance for the mathematical concept of infinite.

References

Balaguer, M. (1998). *Platonism and anti-platonism in mathematics*. Oxford: Oxford University Press.
Balaguer, M.: (2009). Fictionalism, theft, and the story of mathematics. *Philosophia Mathematica, 17*(2), 131–162.
Bernays, P. (1983). On platonism in mathematics. In H. P. Paul Benacerraf (Ed.), *Philosophy of mathematics: selected readings* (pp. 258–271). Cambridge University Press.
Black, M. (1971). The elusiveness of sets. *Review of Metaphysics, 24*, 614–636.
Boolos, G. (1984). To be is to be a value of a variable (or to be some values of some variables). *The Journal of Philosophy, 81*(8), 430–449.
Boolos, G. (1985). Nominalist platonism. *The Philosophical Review, 94*(3), 327–344.
Brown, J. (2003). Kitcher's mathematical naturalism. *Croatian Journal of Philosophy, 3*, 1–20.
Burgess, J. P. (2004). Mathematics and bleak house. *Philosophia Mathematica, 12*(1), 18–36.
Dummett, M. (1963). The philosophical significance of Gödel's theorem. In M. Dummett & P. Tartaglia (Eds.), *Ratio* (pp. 186–214). Duckworth.
Hoffman, S. (2004). Kitcher, ideal agents, and fictionalism. *Philosophia Mathematica, 12*, 3–17.
Kitcher, P. (1984). *The nature of mathematical knowledge*. Oxford: University Press.
Lewis, D. (1993). Mathematics is megethology. *Philosophia Mathematica, 1*(1), 3–23.
Parsons, C. (1990). The structuralist view of mathematical objects. *Synthese, 84*(3), 303–346.
Resnik, M. D. (1988). Second-order logic still wild. *The Journal of philosophy, 85*(2), 75–87.
Thomas, R. (2000). Mathematics and fiction I: Identification. *Logique et Analyse, 43*, 301–340.
Thomas, R. (2002). Mathematics and fiction II: Analogy. *Logique et Analyse, 45*, 185–228.
Velleman, D. J. (1993). Constructivism liberalized. *Philosophical Review, 102*(1), 59.
Yablo, S. (2002). Go figure: A path through fictionalism. *Midwest Studies in Philosophy, 25*, 72–102.

Chapter 3
On Arbitrary Fictional Models

Abstract In this chapter we extend the notion of arbitrary reference to individuals to that of *arbitrary interpretation*. We want to explain how a single *arbitrary* interpretation of the working mathematician relates to the various possible interpretations in model theory. To this purpose we introduce some *arbitrary fictional models*. Additionally, we aim to clarify how one can deduce logical consequences of the axioms by reasoning on a single interpretation, even when a theory has non-equivalent elementary models.

3.1 Introduction

Let L be a first order language. According to model theory, the L-sentences are syntactical entities that become true or false only with reference to a certain interpretation of L. We learn from proof theory how to draw from a given set of sentences logical consequences in full accordance with model-theoretical semantics. According to the soundness and completeness theorems, the sentences formally derivable from a set of assumptions are exactly those true in all possible models of such assumptions.

We want to remark that the soundness and completeness theorems show that formal derivability is sound and adequate not only to the model-theoretical notion of truth, but also to the intuitive notion of *abstract* deducibility, as understood by a working mathematician, whose reasoning is not driven by syntactical inference rules but by an abstract notion of truth. Abstract mathematical reasoning is not concerned with non-interpreted sentences but with genuine propositions that are supposed to be endowed with well-determined truth values relative to an arbitrary interpretation of L.

What has to do the single arbitrary interpretation of the working mathematician with the various possible interpretations of model theory? When a theory has non-equivalent elementary models, how can one deduce logical consequences of the axioms by reasoning on a single interpretation?

M. Carrara, E. Martino, *Arbitrary Reference in Logic and Mathematics*, Synthese Library 490, https://doi.org/10.1007/978-3-031-66452-6_3

3.2 On Epistemic Validity, Epistemic Truth, and Epistemic Completeness

Fictional characters are sometimes conceived of as *incomplete* objects. Is an arbitrary interpretation to be understood as an incomplete thing in a similar sense? Perhaps it would be possible to extend the ontology of ordinary interpretations by adding *incomplete* interpretations, but we think that it would be highly artificial and problematic. In particular, usual mathematical reasoning is in agreement with classical *bivalent* logic, which seems to be inappropriate, when dealing with incomplete objects.

One could think, at first sight, that reasoning about an arbitrary interpretation amounts to reasoning *simultaneously* on *all* possible interpretations. It is not so, however. Abstract reasoning cannot be accounted as referring to all possible interpretations, it genuinely refers to a single interpretation. As we saw, a similar question arises about arbitrary reference to individuals. We want to extend the notion of *arbitrary reference* to individuals, introduced in Chap. 1, to that of *arbitrary interpretation*.

Let us imagine an *ideal mathematician* Kurt and an *ideal agent* Jack. Kurt reasons in agreement with classical logic, never makes any mistake and is free of empirical space-temporal limitations. So Kurt is able to grasp any formal proof and recognize it as sound. The ideal agent Jack is able to fix at will any interpretation of language L. Imagine that Kurt can communicate with Jack. Let Kurt order Jack to choose at his will an interpretation I of L and give him the relevant pieces of information relative to the event of the choice, without communicating him anything about the content of I. So, by merely referring to that act of choice, Kurt can refer to I, without knowing anything about it. We will say that I is a fictional interpretation of L and that the L-sentences, interpreted in I, are fictional propositions. Thus Kurt is able to interpret any L-sentence A in the (fictional) meta-linguistic proposition "A is true in I".

We say that an L-sentence A is *epistemically valid* if Kurt can recognize its truth in any fictional interpretation I. Since Kurt has no piece of information about the content of I, he can recognize A only if A holds in all interpretations, i.e. if it is logically valid. Vice versa, if A is logically valid, it is, by semantic completeness, formally provable. And since Kurt is free of space-temporal limitations, he can grasp such a formal proof of A and recognize the truth of A. Thus:

An L-sentence is logically valid if, and only if, it is *epistemically valid.*

Epistemic validity is easily extendable to epistemic consequence. Let Γ be a set of L-sentences. We say that A is an *epistemic consequence* of Γ if, with reference to a fictional interpretation I of L, Kurt can recognize A under the hypothesis that all sentences in Γ are true in I. Call briefly such hypothesis the *Γ-hypothesis*.

This is nothing but the assumption that I is a model of Γ. Observe that the *Γ-hypothesis* is not to be understood as a condition on Jack's choice. Kurt cannot, in general, command Jack to choose a model of Γ, since Γ might be contradictory and hence devoid of models. Besides, even if Γ is consistent, it is not required, for our purposes, that Jack be able to determine a model of it.

I is an absolutely *arbitrary* interpretation of language L, which Kurt *assumes* to be a model of Γ. More formally, we say that the pair $M = (I, \Gamma)$ is a fictional model of Γ.

So:

A sentence is *epistemically true* (shortly e-true) in the fictional model M if it is an epistemic consequence of Γ.

Of course e-truth in M may fail to coincide with ordinary truth in I, since I may fail, as a matter of fact, to be a model of Γ. The e-true sentences are exactly those recognizable by Kurt as true under the Γ-hypothesis. If Γ is contradictory, every L-sentence is e-true in any fictional model of Γ. Obviously all fictional models are elementary equivalent with respect to e-truth.

From semantic completeness it follows immediately that:

A sentence A is a logical consequence of Γ if, and only if, it is an epistemic consequence of Γ, if, and only if, it is e-true in any fictional model of Γ.

Semantic completeness plays an essential role in the equivalence between logical and epistemic consequence. That may be one of the reasons for preferring first order to higher order logic. For the latter there is no evidence that the notion of logical consequence be adequate, even in principle, to represent what is deducible through abstract reasoning. For an axiomatic first order theory we can translate the semantic completeness theorem into the following

Epistemic completeness theorem. The theorems of an axiomatic first order theory are exactly the epistemic consequences of the axioms, i.e. the sentences e-true in any fictional model of the axioms.

In the next section, we want to exploit the notion of epistemic consequence in order to solve a puzzle arising from the popular understanding of Gödel's second incompleteness theorem.

3.3 The Unprovability of Consistency

Gödel's second incompleteness theorem (Gödel, 1931b) sounds, in its popular version, as follows:

(1) *Popular version of Gödel's theorem.* An axiomatic consistent first order theory, endowing a suitable fragment of arithmetic, cannot prove its own consistency.

This version was formulated by Gödel himself in the famous postscript to his (Gödel, 1931a) paper.

Recall the arithmetical expression of consistency. Let L be the language of first order arithmetic and S a decidable extension (consistent or not) of Peano's axioms PA. Start from the metalinguistic relation that correlates a sentence with a (possible) proof of it. This is expressible arithmetically through a formula $Prf(x, y)$ (read: y is a proof of x) such that:

(i) if n is the code of an S-proof of ϕ, then $Prf(\ulcorner\phi\urcorner, \underline{n})$ is derivable from Peano's axioms PA (where $\ulcorner\phi\urcorner$ is the code of ϕ and $\underline{n}$ the numeral of n);
(ii) if n is not the code of an S-proof of ϕ, then $\neg Prf(\ulcorner\phi\urcorner, \underline{n})$ is derivable from PA.

Define the arithmetical translations of provability and consistency by putting

$$P(x) =_{df.} \exists y Prf(x, y),$$
$$Con =_{df.} \neg P(\ulcorner 0 = 1\urcorner).$$

The precise statement of Gödel's theorem is the following:

(2) If S is consistent, then Con fails to be S-provable.

(1) is the popular reading of (2).

By epistemic completeness, it follows from (2) that, if S is consistent Con fails to be an epistemic consequence of S. So, if $M = (I, S)$ is a fictional model of S, Con fails to be e-true in M. This means that Kurt is not able to recognize the truth of Con in I under the assumption that I is a model of S. Thus, according to the popular reading, we should conclude that Kurt cannot recognize the consistency of S, even under the assumption that S has a model. But that is highly puzzling: Kurt knows that if a theory has a model, then it is consistent! Of course, Kurt doesn't know whether I is really a model of S, but that doesn't matter. The point is that he reasons under the *assumption* that I is a model of S. His consistency proof is the following:

Proof (Simple Proof) The axioms of S are true in I (by the S-assumption), the inference rules are truth-preserving, hence every theorem of S is true in I; no contradiction can be true in I, hence no contradiction is derivable.

Since the soundness of the first order logical rules counts (for a classical mathematician) as a *logical law par excellence*, the *simple proof* seems to be a perfectly logical reasoning from the axioms of S to their consistency. However, the argument to the effect that the consistency of a theory is a trivial consequence of its axioms has often been criticized on the base of a strange objection that we want to challenge.

It is objected that the mere assumption of the axioms does not commit to their truth, since hypothetical reasoning can rightfully assume *any* set of sentences, true or false, consistent or not. So, the objection goes, the assumption of the truth of the axioms is a further assumption. We can use certain assumptions without knowing if they are true; furthermore, the objection continues, hypothetical reasoning is even a powerful means for discovering their possible falsity by investigating their consequences. In this vein it has often been argued that the reason why we can recognize the truth of Gödel's undecidable sentence *Con* of PA is that we, unlike any machine, are very confident that Peano axioms are true, because of our familiarity with arithmetical notions, so that no doubt can arise about their consistency. The formal system, in contrast, cannot exploit the truth of the axioms.

We maintain that such kind of argument rests on a misunderstanding.

Let us consider how a working mathematician reasons. Of course, when she tentatively proposes an axiomatic theory S, she must be well aware of the danger that the axioms may turn out to be contradictory. But, as far as she is drawing consequences from her axioms, she must *pretend* to know that they are true. She reasons abstractly in terms of truth in an arbitrary hypothetical model of S.

In terms of our fictional model, Kurt is well aware that S may be inconsistent, but, when reasoning about the fictional model $M = (I, S)$, he pretends to know that all axioms of S are true in I and that, consequently, their consistency is granted.

However, the objection at issue against the simple proof may seem to be supported by the following argument. If the statement that what is S-provable is true were not a further assumption beyond the axioms, it shouldn't help to prove any S-unprovable L-sentence. However, the argument goes, we can formalize the simple proof as follows.

Extend the language L to a language L' by introducing the truth predicate T for L and extend S to S' by introducing the axiom schemata

(3) $P(\ulcorner \phi \urcorner) \rightarrow T(\ulcorner \phi \urcorner)$,
(4) $T(\ulcorner \phi \urcorner) \rightarrow \phi$ (for any L-sentence ϕ)

We get a non-conservative extension of S, within which *Con* turns out to be a theorem. For, from $P(\ulcorner 0 = 1 \urcorner)$, by using the new schemata, we get $T(\ulcorner 0 = 1 \urcorner)$ and then 0=1, whence, by reduction, *Con*.

So, (3) and (4) certainly do express something not already implicit in the axioms of S.

Moreover, Shapiro in (1998) has observed that, according to the basic intuition of arithmetical truth, it is quite natural to extend the induction principle of PA to T, so that (3) and (4) are formally derivable by induction on the complexity of the proofs. This would suggest, according to Shapiro, that truth is thick and not thin, as the deflationist claims. That seems to refute our main claim that assuming that what is provable is true is not a further assumption beyond the axioms.

It doesn't do to reply that, though (3) and (4) are implicitly granted by the axioms of S, they cannot be exploited in L for the reason that the truth predicate T of L, unlike P, fails to be expressible in L itself. If they were implicitly granted by the S-axioms, so would be the L-schema

(5) $P(\ulcorner 0 = 1 \urcorner) \rightarrow \phi$

as well, so that this would hold in all S-models and, by the completeness theorem, it would be S-derivable and *Con* would be an S-theorem.

Observe, however, that Kurt (deflationist or not) knows, in virtue of the soundness theorem, that under the S-assumptions all the S-theorems are truth in I. But, then, why he is not able to recognize that, under the same assumption, (5) is true in I?

The answer is plain: the arithmetical predicate P may fail to express S-provability in I, even if I is actually a model of S. In fact, I might be a non-standard model of arithmetic, where the truth of $P(\ulcorner\phi\urcorner)$ does not guarantee the provability of ϕ.

Once recognized the consistency of S, Kurt can infer, by clause (ii), that no standard number can satisfy the propositional function $Prf(0 = 1, y)$; however, there remains the possibility that it could be fulfilled by a non-standard number. So Kurt cannot have evidence of $\exists y Prf(0 = 1, y)$, i.e. of Con.

This is the reason why the consistency of S does not imply the e-truth of Con. In other words, what Kurt cannot recognize, when reasoning about his fictional model, is not the consistency of S, but the alleged meaning of Con as the arithmetical translation of consistency.

3.4 On Löb's Theorem

It may be instructive to consider a similar misleading interpretation of

Löb's theorem: $P(\ulcorner\phi\urcorner) \rightarrow \phi$ is provable if, and only if, ϕ is provable.

When reading $P(\ulcorner\phi\urcorner)$ as "ϕ is provable", Löb's theorem seems to suggest that the axioms are inadequate to grant that what is provable from them is true. More precisely, let us extend the language L to the language $L' = L\cup\{T\}$, introducing the truth predicate T for the L-sentences and extending S to S' by adding the Tarskian schema:

(6) $T(\ulcorner\phi\urcorner) \leftrightarrow \phi$ (with $\phi \in L$)

Any model M of S can be extended to a model M' of S' by interpreting T as the truth predicate for M, so that S' is a conservative extension of S. Therefore, if an L-sentence is unprovable from S, then the sentence:

(7) $P(\ulcorner\phi\urcorner) \rightarrow T\phi$

is unprovable from S' (otherwise would be provable from S, against Löb's theorem). And the unprovability of (7) suggests the conclusion that

(8) it is unprovable that what is provable is true.

This conclusion can be criticized by arguing that the mere Tarskian biconditionals are insufficient to characterize truth. Consider, for example, Shapiro's arguments against the deflationist conception of truth (Shapiro, 1998). He observes that, if we expand the language L of PA with the truth predicate T for L, it is natural to extend the induction principle to formulas containing T. In this way one obtains a non-conservative extension of PA, in which the soundness theorem is formalizable and *Con*(PA) is derivable. This would show, according to Shapiro, the deductive power of truth, which would be "thick", not "thin", as the deflationists claim.

Imagine, again, that Kurt is reasoning about the fictional models M of S and M' of S'. Even if, instead of the mere Tarskian schema, he is given the complete information that T is the genuine truth predicate for M, he cannot recognize that induction is applicable to all L'-formulas. In fact, that may be false if M is non-standard.

Turning back to the above alleged suggestion of Löb's theorem, can we rightfully conclude that it is unprovable that what is provable is true?

The correct answer is immediately given by our fictional scenario. Consider, again, Kurt, when reasoning about the fictional models M and M'.

What he cannot know is that,

if $\mathrm{P}(\ulcorner \phi \urcorner)$ is true at M, then ϕ is derivable from S.

In fact, as already observed, M may fail to represent provability at M. Therefore, he cannot know if the sentence

$$P(\ulcorner \phi \urcorner) \rightarrow \mathrm{T}\phi$$

is true at M'. He is well aware, however, that if ϕ is derivable from S, then $\mathrm{T}(\ulcorner \phi \urcorner)$ is true at M'.

At this point, it should be clear that Löb's theorem has nothing to say about the relation between provability and truth. The opposite appearance is caused by the sloppy reading of $P(\ulcorner \phi \urcorner)$ as "ϕ is provable".

P is a faithful translation of provability only in the standard model of arithmetic. But the axioms fail to determine the standard model and therefore the alleged meaning of P as formal provability.

Observe that it has been often argued that the Hilbertian finitary methods of proof are formalizable in PA. Gödel (1931b) cautiously does not endorse this thesis. In the above quoted paper he writes:

> For a system in which all finitary (that is intuitionistically unobjectionable) forms of proof are formalized, a finitary consistency proof, such as the formalists seek, would thus altogether impossible. However, it seems questionable whether one of the system hitherto set up, say Principia mathematica, is so all-embracing (or whether there is a system so all-embracing at all). (Gödel, 1931a, p. 205)

Gödel gives for granted that Con expresses the consistency of S and therefore questions whether there exists a system in which all finitary forms of proof are formalizable. But what we want to hold is that no formal system is adequate even to merely express its own consistency. Gödel's arithmetization of syntactical notions shows that the syntactical structure of a formal system is isomorphic to the standard model of arithmetic. But no formal system can characterize such a model. Hilbert's finitism is grounded on the notion of a finite string of signs, where the attribute "finite" is to be understood in an *absolute* sense. And this *absolute* sense is just what the axioms are not able to catch, because of the impossibility to banish non-standard models.

At this point it should be clear that the popular version of Gödel's theorem is misleading. It gives for granted the reading of the arithmetical predicate P as formal provability and therefore identifies the unprovability of Con with the unprovability of consistency.

However, as we saw, what Gödel's theorem shows is rather that, for any L-predicate P, no system S of axioms can guarantee that P faithfully represents meta-linguistic provability. More precisely, the argument for the unprovability of Con, applied to any L-predicate P, shows that

If S is a consistent extension of PA, there is no L-predicate P such that, for every model M of S and every L-sentence ϕ, $P(\ulcorner\phi\urcorner)$ is true in M if, and only if, ϕ is S-provable.

3.5 Conclusions

We conclude this chapter observing that, despite what is commonly believed, Gödel's theorem has nothing to say about the provability of the genuine consistency of the theory. What it points out is rather a remarkable inadequacy of the axiomatic method, strictly connected with the phenomenon of incompleteness: the impossibility to force, by means of suitable axioms, the material equivalence of an L-predicate with the genuine meta-linguistic provability predicate. The behavior of the latter is not so far from that of the truth predicate as it is usually believed.

We saw hat S-provability is equivalent to e-truth in a fictional model of S. And, if S is consistent, as no L-predicate, interpreted in an ordinary model M of S, can express the truth predicate of M, so no L-predicate, interpreted in a fictional model M of S, can express e-truth, i.e. S-provability, in M.

Finally, we observed that the interpretation of Löb's theorem, according to which it would show that the axioms are inadequate to grant that what is derivable from them is true, is misleading. This interpretation is erroneously suggested by the inappropriate arithmetical translation of the genuine provability predicate.

References

Gödel, K. (1931a). Discussion on providing a foundation for mathematics. In S. Feferman (Ed.), *Kurt Gödel: Collected works* (Vol. 1, pp. 201–205). Oxford: OUP.

Gödel, K. (1931b). On formally undecidable propositions of principia mathematica and related systems. In S. Feferman (Ed.), *Kurt Gödel: Collected works* (Vol. 1, pp. 145–195). Oxford: OUP.

Shapiro, S. (1998). Proof and truth: Through thick and thin. *The Journal of Philosophy*, *95*(10), 493–521.

Chapter 4
On the Logicality of Second-Order Logic in Terms of Plural Arbitrary Reference

Abstract The aim of this chapter is to argue that: (a) our semantics of acts of choices (**SAC**), as developed in Chap. 2, defends second-order logic from claims of ontological commitment; (b) understanding our semantics does not require any prior mathematical concepts; and (c) although **SAC** is not universally applicable, it still offers significant applicability, especially in mathematics. We conclude the chapter arguing that second-order logic, as interpreted through our semantics, can indeed be considered a genuine logic. One might object that, since only concrete objects can be referred to by arbitrary acts of choice, **SAC** lacks the logical requirement of universal applicability. However, such limitation is superseded by the structuralist conception of mathematics, according to which arbitrary entities, independently of their very nature, can play the role of mathematical objects.

4.1 Introduction

As we embark on a deeper exploration of logical systems, we enter a realm where the boundaries blur between the pristine clarity of first-order logic and the more expressive yet complex second-order logic. It is a domain that beckons with the promise of greater expressive power, where quantification extends beyond mere individuals to properties and relations themselves.

This leap in expressive richness, however, does not come without its costs and controversies. The debate at the heart of this discussion centers around the philosophical and mathematical implications of second-order logic's capabilities. Some argue that it transcends the realm of 'pure' logic, encroaching on the territories of set theory and metaphysics, due to its reliance on set-theoretic notions for semantic interpretations.

M. Carrara, E. Martino, *Arbitrary Reference in Logic and Mathematics*, Synthese Library 490, https://doi.org/10.1007/978-3-031-66452-6_4

With this in mind, we confront a pivotal question that cuts to the core of logical theory:

Question: Is second-order logic pure logic?

As observed in Chap. 2, according to Quine, second-order logic is set theory in disguise:[1] in the light of his slogan *to be is to be the value of a (bound) variable*, quantification on second-order variables entails that second-order logic is committed to the existence of sets as second-order entities, since sets, unlike intensional entities, enjoy an identity criterion.[2] Boolos—as analysed in Chap. 2—proposed in (1984) and (1985) a way to interpret monadic second-order logic without ontological commitment to second-order entities. According to his proposal, second-order variables do not vary over classes or properties, rather they vary *plurally* over the individuals of the domain of discourse.

Linnebo in (2003) argues that Boolos' plural interpretation does not vindicate second-order logic from Quine's allegation. He provides three requirements that a formal theory should satisfy in order to be considered as logic, i.e. *ontological innocence*, *universal applicability*, and *cognitive primacy*.

In general, logic should not be committed to the existence of any kind of entities; it should be applicable to any domain of discourse; and its understanding should be direct (i.e. not involving mathematical notions as set theory or combinatorics). On the basis of these criteria, in (2003) Linnebo argues that Boolos' plural quantification does not account as *pure logic*.

Aim of the chapter is to argue that (a) our semantics of acts of choices (**SAC**), as it has been developed in Chap. 2, vindicates second-order logic from the allegation of being ontologically committal; (b) that the understanding of our semantics does not presuppose any mathematical notion; and that, (c) though, strictly speaking, not universally applicable, **SAC** still provides a fair amount of applicability, especially to mathematics.

We will conclude that second-order logic, as interpreted by our semantics, can be considered as pure logic, and that the only limitation it suffers from, i.e. a lack of universal applicability, is a minor drawback.

4.2 Linnebo's Objections to Boolos' Plural Quantification

In (2003) Linnebo argues that any account of the logicality of plural logic comes down to providing grounds for the logicality of the *plural comprehension principle* (**CP**). And, since the notion of *arbitrary sub-plurality* relies on some combinatorial and maybe even set-theoretical intuition, the plural comprehension principle can

[1] The correct quotation is 'set theory in sheep's clothing': Quine (1986), ch. 5.

[2] On this see Carrara and Giaretta (2004).

hardly be considered a principle of pure logic. Consequently, he maintains, Boolos' argument for the logicality of plural logic fails.

Let us see Linnebo's argument in detail.

According to Linnebo, in order for the theory of plural quantification to be considered as a logical theory, (at least) the following conditions should hold:

Ontological Innocence. The plural comprehension axioms are not ontologically committed to any entities beyond those already accepted in the ordinary first-order domain;

Universal Applicability. The theory of plural quantification can be applied to any realm of discourse in any understanding of mathematical ideas, no matter what objects this discourse is concerned with;

Cognitive Primacy. The theory of plural quantification presupposes no mathematical ideas in order to be understood, but can be understood directly. Our understanding of it does not consist, even in part, in an understanding of mathematical ideas, such as ideas from set theory or from other branches of mathematics.[3]

Linnebo claims that, for the logicality of plural quantification to be ensured, the logical status of the plural comprehension principle is to be investigated. This involves the notion of an *arbitrary sub-plurality* of individuals. According to Linnebo (2003), the notion of an arbitrary sub-plurality hinges on combinatorial—and maybe even set-theoretical—concepts. An arbitrary sub-plurality is the output of a combinatorial operation on individuals, that one can perform by e.g. going through them and 'selecting some and rejecting others'.[4] So, clearly, we need some, though elementary, understanding of combinatorial operations, in order to understand the very notion of sub-plurality. Furthermore, according to him the endorsement of such combinatorial notions would commit us to allow for *collecting together* in turn the pluralities so construed and forming higher-order pluralities. If so, we would be led by the plural interpretation of second-order logic to larger and larger extensions of the domain of individuals. So, if second-order logic were pure logic in virtue of plural quantification, such would be also higher- and higher-order logics.[5] Linnebo concludes that the plural interpretation fails to make the impredicative comprehension principle a genuine logical principle. In fact, allowing for the iteration of the operation of sub-plurality leads Boolos' plural logic to the inexhaustibility of the layers of higher and higher order pluralities of some combinatorial ideas. We could not account, then, for the universal applicability of plural logic, since it would not make us able to talk about all the pluralities there are. Also, the appeal to combinatorics and set theory shows, according to Linnebo, that Boolos' plural logic is not even cognitively prior to mathematics, since the

[3] Linnebo (2003, p. 77).

[4] Linnebo (2003, p. 86).

[5] See Parsons (1990).

understanding of the notion of arbitrary sub-plurality requires the understanding of some mathematical ideas.

A short paragraph of Linnebo (2003) clarifies his position on pluralities:

> [A defense of Boolos' position] is based on the idea that only *things* can be collected together. If this idea is right, and if Boolos avoids reifying pluralities, then there will simply be no *things* available to be collected together to form higher pluralities. But this defense too is unconvincing. There is no obstacle to iterating the combinatorial considerations that give content to our talk about arbitrary sub-pluralities; in particular, combinatorics has no ontological qualms about collecting together first-order pluralities so as to form higher pluralities. For instance, from the point of view of combinatorics, it is no more problematic to arrange individual Cheerios in the following way: 00 00 00 than it is to arrange them as: 000000, although the former arrangement is most informatively described as three pairs of Cheerios x which is a higher order plurality x whereas the latter arrangement is a mere first-order plurality based on the same six Cheerios. To whatever extent the more complex arrangement involves additional ontological commitments, these commitments pose no problem to combinatorics. (Linnebo, 2003, pp. 87–88)

Observe that, according to such alleged consequence, combinatorics would be a very interesting powerful means for generating higher and higher mathematical infinities from any given infinity of individuals without introducing new entities at all.

It is clear from this passage that, according to Linnebo, avoiding to reify pluralities does not exclude that, after all, first-order pluralities are capable to be collected together. That combinatorics has no qualms about collecting together first-order pluralities comes down that it is completely indifferent to the ontological status of such first-order pluralities.

As the example of cheerios shows, higher-order pluralities are simply produced, according to Linnebo, by a certain mode of collecting first-order individuals.

Observe that, once accepted Linnebo's understanding of combinatorics one could see it as a mirabilis powerful tool for generating from any given plurality of individuals higher and higher mathematical infinities, without the need of new entities at all.

On the contrary, we want to defend Boolos' thesis that plural quantification of first-order individuals by no means involves the possibility of higher-order quantification.

4.3 Is Second-Order Logic Logic at All?

Once accepted Linnebo's requirements, in order to look into whether a theory T is pure logic, T's semantics has to be investigated concerning its own *ontological commitment*, *universal applicability*, and *cognitive primacy*.

Since we have proposed the *semantics of acts of choice* (**SAC**) as an alternative to Boolos' semantics, we have to discuss **SAC** with respect to the three aforementioned conditions.

Firstly, consider ontological innocence.

Question: Is **SAC** committed to any further entities beyond those of the first-order domain?

First of all, it is clear from Chap. 2, Sect. 2.4 that a *p.s.c.* does not create any entity that collects the chosen individuals. Speaking of pluralities as if they were genuine entities is a mere *façon de parler*, paraphrasable in terms of plural choices. Thus, plural reference does not involve the notion of class. **SAC**, in fact, substitutes the notion of assignment function/relation with the notion of act of (arbitrary) choice, where acts are not entities at all.

One may wonder if the use of the team of agents counts as an ontological commitment beyond the first order domain **D** of individuals. In Chap. 1 we have argued that arbitrary reference is essential both to formal and informal reasoning, so that it counts as a logical notion. Plural arbitrary reference is nothing but a natural generalization of that. The team of agents with their acts of choice are used as a device for explaining plural arbitrary reference. In this sense they are strictly functional to logic. Besides the cardinality of the team of agents is the same as that of the first order domain of quantification, so that, unlike classical set-theoretical semantics for second order logic, no greater infinity is involved by **SAC**. For these reasons we think that **SAC** satisfies the requirement of ontological innocence and cognitive primacy.

Consider universal applicability.

Question: Can **SAC** be applied to any realm of discourse, no matter what objects this discourse is concerned with?

As explained in Chap. 1, **SAC** applies to a domain of concrete individuals. To be concrete is essential to be an object of direct reference by ostension. In this respect, if one is open to admitting the existence of abstract entities, then she will object that **SAC** fails to be universally applicable. We indeed concede this objection. However, it is not necessary to appeal to the existence of abstract entities in order to make sense of mathematics.

In fact, the structuralist view of mathematics shows how the whole mathematics can be developed assuming the existence of a suitable infinity of individuals of whatever nature (see, for example, Lewis' *megethology* developed in *Mathematics is megethology* (Lewis, 1993)). So, one can assume, in particular that such individuals are concrete. Thus, the concreteness of individuals does not undermine the generality of mathematical reasoning, To this extent the requirement of *universal applicability* is not essential to the development of mathematics.

Observe that the agents, though imaginary characters, are thought of as concrete human beings, so that they can play both the role of choosers and that of chosen.

Consider, finally, cognitive primacy.

Question: Does **SAC** presuppose any mathematical ideas to be understood, such as ideas from set theory or any other branches of mathematics?

As mentioned, **SAC** does not rely on the understanding of mathematical notions, e.g. functions or relations, unlike standard semantics for second-order logic.

However, it might be disputable whether the notions by which we understand **SAC** can be considered as legitimate part of the logician's toolkit.

SAC serves the purpose of accounting for reference in formal languages. Nonetheless, though the notion of reference is not *defined* in **SAC**, it is *elucidated* in terms of acts of choice. Our understanding of it amounts to the understanding of the crucial notions which **SAC** revolves around.

In this respect, the issue of the cognitive primacy is twofold. First of all, it concerns the logical status of (arbitrary) reference. Secondly, since **SAC** is tightly connected to the notion of infinitely many simultaneous acts of choice, the role of the notion of infinity has to be taken into consideration.

About the first point, observe that in general we think that the appreciation of the logicality of a formal theory is not separated from the appreciation of its semantics. Thus, the notion of (arbitrary) reference has to be considered as a part of logic, since otherwise we would end up with sets of uninterpreted symbols whose ontological commitment, universal applicability, and cognitive primacy could not be appreciated at all.

One can argue that if first-order logic counts as genuine logic, second-order logic, interpreted through **SAC**, counts as genuine logic too in the following way. Take the notion of reference: as said it is essential to first-order logic. Now consider our acts of choice. As we said, they are not entities at all. On the other hand, agents are obviously entities. So one may wonder if our team of agents, possibly infinite, is to be regarded as an extra-logical idealization. Observe, however, that our agents play the role of performers of acts of choice for referring to individuals of the first-order domain. Besides, they are as many as such individuals. Their activity is nothing but an idealization of the usual actions of reference of working mathematicians. So, as far as singular reference is regarded as a logical notion, it seems quite natural to extend logicality to plural reference (on this point see also Sect. 7.6).

Moreover, one can observe that the notion of (arbitrary) reference explained in terms of acts of choice is cognitively more intuitive than the notion of a function or relation in a formal setting. We do not need a mathematically sophisticated apparatus to understand what it means to choose something over something else. The notion of act is not mathematical at all. An appreciation of it naturally comes from everyday life, when we constantly perform acts by which we choose things.

Finally, the fact that *infinitely* many simultaneous acts of choice may be appealed to in order to elucidate reference to pluralities does not seem to be problematic. Our agents function as performers carrying out acts of choice referring to individuals within the first-order domain, and their number is the same of that of first-order individuals.

To this extent, second-order logic interpreted through our **SAC** with infinitely many choices is no less logic than first-order logic.

4.4 Conclusions

In this chapter, we considered the criticisms moved against Boolos' plural semantics for second-order logic, in particular Parsons' and Linnebo's criticisms. Linnebo's criticisms pertain to ontological innocence, universal applicability, and cognitive primacy. We analysed them.

We claimed that our semantics for second-order logic is ontologically innocent because it does not presuppose any entity outside the first-order domain. Besides, **SAC** does not involve extra-logical notions insofar as the team of agents has been introduced to the purpose of explaining the logical notion of reference. Finally, as to *cognitive primacy*, this seems to be in agreement with the primitive intelligibility of the notion of act of choice used as a device for explaining logical inference.

References

Boolos, G. (1984). To be is to be a value of a variable (or to be some values of some variables). *The Journal of Philosophy*, *81*(8), 430–449.

Boolos, G. (1985). Nominalist platonism. *The Philosophical Review*, *94*(3), 327–344.

Carrara, M., & Giaretta, P. (2004). The many facets of identity criteria. *Dialectica*, *58*, 221–232.

Lewis, D. (1993). Mathematics is megethology. *Philosophia Mathematica*, *1*(1), 3–23.

Linnebo, Ø. (2003). Plural quantification exposed. *Noûs*, *37*(1), 71–92.

Parsons, C. (1990). The structuralist view of mathematical objects. *Synthese*, *84*(3), 303–346.

Quine, W. (1986). *Philosophy of logic*. Harvard: Harvard University Press.

Chapter 5
A Notion of *Logical Concept* Based on Plural Arbitrary Reference

Abstract In this chapter, building on the previous chapters' approach to plural quantification through plural arbitrary reference grounded on the semantics of plural acts of choice, we develop a theory of concepts termed as a theory of *logical concepts*. Within this framework, we propose a novel logicist approach to natural numbers.

5.1 Introduction

In this chapter, using our theory of plural arbitrary reference and the associate semantics (**SAC**), introduced in Chap. 2, we develop a predicative theory of *logical concepts* (**LC**). The leading idea is that logical concepts are definable in a logical language and have as range of application the domain of their linguistic expressions.

To this purpose we will use a device introduced by Nino Cocchiarella in Cocchiarella (1985) and Cocchiarella (1986). He adopted a language of second order logic with nominalization for revising Frege's ill-famed *Law V*, according to which every concept is correlated with an object. In such a language second order variables range both over concepts and objects. A second order variable can occur both in predicate and in subject position: when occurring in a formula both in subject and predicate position, it represents, in subject position, the object correlated with the concept that it represents in predicate position. Similarly, our concept variables will range over our closed abstracts both in their roles of concepts and of mere strings of signs. Their positions, in a formula, will distinguish the two roles.

To summarize: we develop a theory of logical concepts that leads to a new logicist theory of natural numbers. The logicist flavor of the theory is discussed in the last section of the chapter.

M. Carrara, E. Martino, *Arbitrary Reference in Logic and Mathematics*,
Synthese Library 490, https://doi.org/10.1007/978-3-031-66452-6_5

5.2 The Formal Language

We introduce the following language with three sorts of variables:

Variables
$x, y, z \ldots$ singular individual variables;
$X, Y, Z \ldots$ plural individual variables;
$F, G, H \ldots$ singular concept variables.

Logical Constants

1. ε: a relation of form "$b\varepsilon X$" to be read as: "b is one of the Xs".

We prefer to use in the sequel the notation "$b\varepsilon X$" instead of "Xb", in order to reserve juxtaposition for predication. Indeed, when intepreted in terms of plural quantification, *second order logic* could be more properly called *plural first order logic*.

2. The usual connectives ($\neg$, $\vee$, $\wedge$, $\rightarrow$, $\leftrightarrow$) and quantifiers ($\exists$, $\forall$) for all variables.

Terms and Formulas

We define *individual terms*, *concept terms* and *formulas* by a simultaneous induction:

(a) Each term is an individual term (concept terms included);
(b) each singular individual variable is a term;
(c) each concept variable is a concept (and individual) term;
(d) if A is a concept term and b is an arbitrary term, Ab is a formula;
(e) if X is a plural (individual) variable and b is an arbitrary term, $b\varepsilon X$ is a formula;
(f) if A, B are formulas, $\bullet$ is a binary connective, and v is an arbitrary variable, $\neg A$, $A \bullet B$, $\exists v A$, $\forall v A$ are formulas;
(g) if A is a formula without quantified concept variables and free plural variables, and x is a singular individual variable, then $[x\colon A]$ is a concept term. Such a term is said an *abstract*: x and A are, respectively, the *defining variable* and the *defining formula* of the abstract. The abstract is *closed* if x is the only free variable of A.

So a concept term is a concept variable or an abstract: it can occur in a formula in subject position or in predicate position. A singular individual variable is an individual term but not a concept term: it cannot occur in predicate position.

A *bound occurrence* of a variable in a term or in a formula is either (i) a quantified occurrence or (ii) a defining occurrence of an abstract. (An occurrence of) a variable is *free* if it is not bound. A term or a formula *is closed iff* all its variables are bound. The restrictions imposed to the defining formulas of the abstracts will be motivated in the description of the intended model in the next section.

Identity

Identity is defined as follows:

$$\text{if } a,\ b \text{ are terms, then } a = b =_{df.} \forall X(a\varepsilon X \leftrightarrow b\varepsilon X).$$

5.3 The Intended Model

The domain D of individuals is constituted by all *closed abstracts*.

Two *abstracts* are regarded as the same entity when they are typographically identical up to *alphabetic variants* (i.e. when one is obtained from the other by a change of *bound variables*).

The abstracts play both the role of individuals, mere linguistic expressions, and of the concepts they express.

A *closed* abstract $[x\colon A]$ is regarded as the concept of all individuals satisfying the formula A.

The *satisfaction* relation will be determined in the next section by an inductive *truth* definition on the complexity of a sentence.

According to these lines, the linguistic use of individual (singular and plural) variables is different from that of the concept variables. A singular individual variable cannot occur in predicate position, because its intended range is the domain of the closed abstracts, understood as mere syntactical entities, which do not play any predicative role. Similarly, no predicative role is played by plural variables: a formula as $t\varepsilon X$ means that t is one of the Xs chosen by a certain plural act of choice, performed by a team of countably many agents (as many as the closed abstracts).

The restrictions imposed to the *abstracts* are motivated by a definitional predicative conception of logical concept. The definition of a *concept* is understood as *constitutive* of the concept, so no quantification over all concepts (on pain of vicious circularity) is allowed in the *defining formula* of an abstract. In contrast, in such a formula, concept variables and singular individual variables can occur free: an abstract with such variables is a schema that yields a concept for every substitution of them with closed abstracts.

On the other hand, it is possible to use singular and plural quantification on individuals because, as purely syntactical objects, they are completely determined by the syntactical formation rules. We allow that concept variables occur also in subject position in order to express in the formal language the fact that any individual defines a concept and any concept is defined by an individual. We adopt, therefore, the following principles:

$$\forall x \exists F(F = x)$$

and:

$$\forall F \exists x(F = x).$$

In this way one can express the fact that the concept defined by x applies to y:

$$\exists F(F = x \wedge Fy).$$

Observe, however, that such a formula, because of the quantification over all concepts, cannot be the *defining formula* of an abstract: it has a mere *descriptive* (and not *constitutive*) role. This explains the reason of the restrictions for concept variables in the defining formula of an abstract.

As to the restrictions to plural variables, these cannot occur free in an abstract, because we do not suppose that any plurality of individuals can be selected by a concept. In our perspective a concept must be linguistically determined, so that it cannot depend on the outcomes of an arbitrary plural choice. Observe that, if the formula $x \varepsilon X$ were a defining formula of an abstract, then every plurality would be the extension of a concept. In fact, in this case, the abstract $[x\colon x \varepsilon X]$ would yield a concept for every *p.s.c.* and one would easily reproduce Russell's paradox.

In contrast, the occurrence in an abstract of a free singular variable is allowed because such abstract is a schema of the concepts, linguistically definable, obtained by replacing the variable with a closed abstract. Besides, a plural variable is allowed to occur quantified in the defining formula of an abstract. In fact, while the language cannot be able to describe the individuals selected by a specific plural choice, nothing prevents the linguistic construction of a concept from using locutions of kind "however certain individuals are chosen..." or "it is possible to perform a choice such that...."

5.4 Formal Semantics

From here on we particularize the above semantics by taking as individuals the closed abstracts of the language so that every individual is definable in the language. An *assignment* a for a formula A assigns to each singular (individual or concept) free variable of A a closed abstract, to each plural free variable of A none, one or more (even infinitely many) closed abstracts. If a does not assign any abstract to X, this variable represents the *empty plurality*. So it is clear that an assignment can be realized by performing a singular or a plural choice for each free variable of A.

Let us define the *truth-conditions* for a formula A relative to an assignment a:

(a) suppose, for the moment, that A has no concept variables and proceed by induction on the complexity of A:

1. $A := t \varepsilon X$. A is true if the term t^*, obtained from t by substituting its free variables with the terms assigned to them by a, is, up to alphabetic variants, one of the individuals assigned to X by a.
2. $A := [x : B]t$. A is true if $B[t/x]$ is true.
3. The truth conditions for the propositional connectives ($\neg, \vee, \wedge, \rightarrow, \leftrightarrow$) are as usual.

4. $A := \exists v B$ (where v is a singular or plural individual variable). A is true if a can be extended to an assignment a^* for B such that B is true.

(b) Let A be an arbitrary formula with concept variables. We proceed by induction on the number of concept variables:

1. A is a formula without quantified concept variables. A is true if the formula A^*, obtained by substituting to the free concept variables the closed terms assigned by a, is true. Notice that A^* has no concept variables.
2. Usual clauses for propositional connectives ($\neg$, $\vee$, $\wedge$, $\rightarrow$, $\leftrightarrow$).
3. $A := \exists F B$. A is true if there is a closed term t such that $B[t/F]$ is true.
4. $A := \forall F B$. A is true if, for all closed term t, $B[t/F]$ is true.

5.4.1 Logical Principles

Let us introduce the crucial logical principles of our theory. The plural comprehension principle (**PCP***)

$$(\mathbf{PCP*})\ \exists X \forall y (y \varepsilon X \leftrightarrow A)\ (X \text{ not free in } A).$$

holds for pluralities of individuals, as already observed in Sect. 5.4.

As to concepts, the following restricted comphrehension principle holds:

(**SCP**) (*Singular comphrehension principle*) $\exists F(F = t)$, where t is a term.

It is satisfied in the intended model because, for any assignment to the free variables of t, this becomes a closed abstract that defines a concept. It is also evident the principle

$$\exists x(x = t).$$

The *conversion principle*

$$(\mathbf{Conv})\ [x:\ A]t \leftrightarrow A[t/x]$$

holds in the intended model in virtue of clause (a2) of the truth definition. One can reformulate (**SCP**) in the form:

$$(\mathbf{SCP'})\ \exists F \forall y (Fy \leftrightarrow A)\ (F \text{ not free in } A)$$

where A is without quantified concept variables and free plural variables. For, these restrictions on A are the same required for the defining formula of an abstract. Since closed abstracts play the double role of concepts and individuals, it is evident the following Frege's principle:

$$\textbf{(FP)}\ \forall F \exists x (x = F).$$

In the present theory **FP** is the *intensional* counterpart of the famous *Law V*. As is well-known, it is debatable whether the original *Law V* counts as a logical principle; furthermore, there is a large consensus that *Law V* is responsible for Russell's paradox. In fact, it is not clear at all the nature of the Fregean correlation between concepts and objects. Frege himself, once he knew the paradox, expressed some doubts on *Law V*.

In contrast, in our present system, **FP** is, as already observed, a self-evident principle: the relation between concepts and individuals is the mere identity. Unlike Frege's *Law V*, **FP** is intensional: it does not require that the correlates of coextensive concepts are identical individuals. In fact, in the intended model, two distinct abstracts may play the role of co-extensive concepts. For instance, consider the abstracts $[x\colon x = x]$ and $[x\colon x = x \wedge x = x]$: since they are not typographical identical (up to alphabetic variants) they are distinct individuals, while as concepts they are coextensive.

Assume the usual principles of identity (*reflexivity* and *substitutivity*). These trivially hold for the typographical identity (up to alphabetic variants). Define a singleton $[x]$ and an ordered pair $[x, y]$ by putting

$$[x] = [z : z = x]$$

$$[x, y] =_{df.} [z : z = x \vee z = y]$$

Assume the axiom

$$[x, y] = [x', y'] \leftrightarrow x = x' \wedge y = y'.$$

Again, this trivially holds for the intended typographical identity.

5.5 The Consistency of the Theory

One can easily show that our intended model is interpretable in the standard model $\mathbb{N}$ of classical second order Peano arithmetic PA.

Let us introduce a numerical codification of our language. We choose the set K of codes of the closed abstracts in such a way that two abstracts have the same code *iff* they differ at most for alphabetic variants.

Then the inductive truth definition for the sentences is expressible in PA. For the purpose, let the variables X, Y,... range over the subsets of K and the variables x, y,... F, G,... range over the members of K. Then take as an assignment any second order binary relation between the set $V \subset N$ of (the codes of) variables and K that assigns a unique member of K to every singular variable and without any restriction on plural variables. The true sentences are the members of the least subset T of the set of (the codes of) the sentences satisfying all clauses (a) and (b) of Sect. 5.3.

Now, observe that the logical principles in Sect. 5.4 are satisfied by the above truth definition, so that they are consistent with PA. The *plural comprehension principle* **PCP*** becomes the classical second order comprehension principle endorsed in PA. **SCP** and **FP** are trivial because both variables x and F range over the same set K. The conversion principle **Conv** holds by clause (a2.)

$$[x : B]t \text{ is true } \textit{iff } B[t/x] \text{ is true.}$$

SCP' holds because the restrictions on A are the same required for the defining formula of an abstract.

The principles of identity and the axiom of ordered pairs hold because of the above conditions on the codes of the closed abstracts.

5.6 Relations and Sets

In our system **LC** we can express two kinds of relations: *plural relations*, generated by plural relating choices *p.r.c.* and *singular relations,* i.e. concepts of ordered pairs. An important singular relation is the relation η defined as follows:

$$x\eta y =_{df.} \exists F(F = y \wedge Fx) \text{ (to be read: “}x\text{ falls under }y\text{” or “}y\text{ applies to }x\text{”).}$$

As intensional entities, concepts do not satisfy the extensionality principle: coextensive concepts may not be identical. However, we will define in the next section an extensional identity between concepts, with respect to which they can play the role of sets.

The extensional identity for concepts ($\cong$) is inductively definable by means of the following clauses:

(a) If x, y are intensionally identical ($x = y$), then they are extensionally identical ($x \cong y$);
(b) if x, y satisfy the condition that every individual falling under one of them is extensionally identical to an individual falling under the other, then $x \cong y$.

Such inductive definition is formalizable in terms of plural relations:

$$x \cong y =_{df.} \forall X(\forall u \forall v(u = v \vee (\forall z(z\eta u \rightarrow \exists w([z, w]\varepsilon X \wedge w\eta v)) \\ \wedge \forall z(z\eta v \rightarrow \exists w([z, w]\varepsilon X \wedge w\eta u)) \rightarrow [u, v]\varepsilon X) \rightarrow [x, y]\varepsilon X)$$

(in words: the extensional identity is the least equivalence relation correlating two individual x and y provided that it correlates every individual falling under one of them with an individual falling under the other). Then, we define the membership relation as follows:

$$x \in y =_{df.} \exists z(z \cong x \wedge z\eta y).$$

With respect to the defined extensional identity and membership our individuals form a set-theoretical structure: in this role we will call them *logical sets*. They satisfy the extensional principle:

(EXT) *Two sets are identical iff their members are identical* (as sets).

A formula will be said to be set-theoretic if it is expressible in terms of the membership relation. In this way, we obtain the *logicist fragment* of pure set theory. In particular, this fragment includes the *hereditarily finite sets* (*h.f.s.*), inductively defined by the following clauses:

(i) $\emptyset = [x \colon x \neq x]$ is a *h.f.s.*;
(ii) if $[x \colon A]$ and $[x \colon B]$ are *h.f.s.*, so are $[x \colon A \vee B]$ and $[x \colon A \wedge B]$;
(iii) if u is a *h.f.s.*, so is $[u]$.

5.7 Natural Numbers

As is well known, arithmetic is interpretable in the theory of *h.f.s.* For example, one can define the set $\mathbb{N}$ of natural numbers à la Zermelo, identifying the successor operation with the singleton operation:

$0 =_{df} \emptyset$,
$Sx =_{df} [x]$,
$\mathbb{N} =_{df.} [x : \forall X(0 \in X \wedge \forall y(y \in X \rightarrow Sy \in X) \rightarrow x \in X)]$.

By using plural relations, one can easily reformulate the inductive definitions of the usual arithmetical operations and prove second order Peano axioms.

So we obtain a theory of natural numbers that may be considered a form of structural logicism, insofar as it is grounded on the theory of logical concepts. We will make later some further comments on this form of logicism.

5.8 Russell's Paradox

Since **LC** is consistent no paradox can arise. It may be instructive, however, to stress how Russell paradox is avoided. Russell set-theoretical paradox does not follow because the formula:

$$\neg x \in x$$

cannot occur in an abstract, since membership $\in$ is defined by means of quantification over concept variables.

As to the intensional version of Russell's paradox, concerning the alleged concept of all concepts that are not falling under themselves, observe that the condition of not falling under itself $\neg(x\eta x)$ is expressed by the formula:

$$\exists F(x = F \wedge \neg Fx),$$

which, again, because of the existential quantification over concept variables, cannot be the defining formula of an abstract. Observe that, in virtue of **(PCP)***

$$\exists X \forall x(x\eta X) \textit{ iff } \exists F(x = F \wedge \neg Fx)$$

So, if every plurality were the extension of a logical concept one could reproduce Russell's paradox.

However, in our theory the alleged concept of all individuals satisfying the formula $\exists F(x = F \wedge \neg Fx)$ does not count as a logical concept in accordance with our predicative understanding of logical concepts.

5.9 The Logicist Flavor of the Theory of Logical Concepts

Our theory of logical concepts developed in this chapter can be regarded as a form of logicism. We want to emphasize, however, that our form of logicism is not to be understood according to the Fregean view, which aims to supply a logical definition of natural numbers in terms of the *general* notions of *concept* and *object*.

Frege's foundation of arithmetic can be labeled as "logicist" because, *inter alia*, it exploits only very general aspects of such notions and generality is usually considered as a peculiar feature of logic. However, his theory quantifies over *all* concepts and *all* objects, so that its ontological commitment, according to Quine's criterion, is not restricted to any sort of logical entities (in any reasonable sense of the word "logical"), but includes an enormous infinity of entities of any nature. In this respect, Frege's numbers are not defined in terms of *logical* entities.

On the contrary, our interpretation of second-order logic—using *acts of choice* instead of *second order entities*—allows a strong reduction of the ontology involved in a logicist foundation of natural numbers. Our *abstracts*, the individuals we quantify over, are countably many types of finite strings of signs. Such entities, called by Parsons *quasi-concrete objects* (as types of concrete entities), are capable of being chosen by *ostension*, as required by our theory of plural choices. Furthermore, they are very basic objects and the human ability to grasp them is essential to the development of logic, as was clearly emphasized by Hilbert in the following passage:

> [A]s a condition for the use of logical inferences and performance of logical operations, something must already be given to our faculties of representation, certain extralogical concrete objects that are intuitively present as immediate experience prior to all thought. If logical inference is to be reliable, it must be possible to survey these objects completely

> in all their parts, and the fact that they occur, that they differ from one another, and that they follow each other, or are concatenated, is immediately given intuitively, together with the objects, as something that neither can be reduced to anything else nor requires reduction. This is the basic philosophical position that I consider requisite for mathematics and, in general, for all scientific thinking, understanding, and communication. And in mathematics, in particular, what we consider is the concrete signs themselves, whose shape, according to the conception we have adopted, is immediately clear and recognizable. (Hilbert 2002, p. 376)

In this sense our abstracts, though *extra-logical* entities, can be labelled as *pre-logical* entities. Besides, as we saw, they can play the role of certain particular concepts with respect to a suitable relation of application. We want to enlighten a sense according to which such concepts can be labeled as *logical concepts*.

In virtue of the restrictions imposed to the abstracts, the application relation among abstracts can be inductively defined, by means of logical connectives and quantifiers, in terms of the syntactical identity among abstracts, considered as mere strings of signs. Individual identity, though formally definable in terms of plural quantification, is essentially a *primitive relation*. Indeed, it is implicitly presupposed in the very same notion of individual, and hence in *singular* and *plural quantification* over individuals. For this reason identity is usually regarded as a *logical relation*. So the concepts represented by our abstracts are in turn of logical nature, insofar as they are logically defined in terms of a logical relation through logical constants. In this sense, the Hilbert's extra-logical entities (finite strings of signs) can be structured as *logical concepts*. Similarly for their role as logical sets (i.e. extensions of logical concepts). And our approach to number theory can be regarded as a form of *logicism*, insofar as all arithmetical notions are definable within our theory of logical concepts.

It may be worthwhile to compare our logicism with *neologicism* (for an introduction to neologicism see Linsky and Zalta 2006). We just sketch here the most relevant features for our purposes. Second-order arithmetic is derivable in second-order logic from what is usually referred to as Hume's Principle:

(HP) $\#F = \#G$ *iff* F and G are equinumerous.

In the standard interpretation F and G range over second-order entities (as concepts or classes), while # is a function mapping second-order entities onto individuals. Neologicism attempts to revise Frege's logicism by replacing Frege's ill-famed *Basic Law V* with **HP**. In fact, Frege exploits Law V only to derive **HP**. The latter is of the same kind as Law V but, unlike Law V, it turns out to be consistent with classical second-order arithmetic. Crispin Wright and other neologicists hold the highly controversial thesis that **HP** counts as a logical definition of numbers (see for example Wright 1983, 1992). See also Linsky and Zalta (2006).

The main argument against this thesis is that **HP** has ontological consequences: the existence of infinitely many numbers. But, following Boolos, it is common to argue that pure logic should have no ontological consequences. If so, there is no hope to derive the existence of numbers from logical principles: arithmetic requires infinitely many individuals, whose existence cannot be guaranteed by logic. Besides, **HP** has been charged to be in "bad company", because of its similarity with Frege's inconsistent *Basic Law V*.

In contrast, our logicism is compatible with the thesis that the existence of no entity can be granted by mere logical means. In fact, we do not claim that the existence of the entities we are committed to is guaranteed by logic. What we claim is that such entities are of a peculiar logical significance insofar as they are functional to the development of logic. In fact, our abstracts as syntactical entities are pre-logical in the explained sense.

Concerning the ontological commitment we think, unlike Quine, that a theory is committed not only to the entities it quantifies over, but to all entities assumed in the object language and in the meta-language. Accordingly, our ontological commitment consists both of the syntactical entities, including the close abstracts (our individuals) and of the agents (the entities we appeal to in the meta-language).

Following Hilbert we have argued for the logical relevance of finite strings of signs. Besides our abstracts are structured, by means of a plural logic, in such a way to play, in turn, the roles of concepts, sets, and numbers.

As to our agents, they are, of course, fictional characters, so in this respect, our philosophical perspective is a form of *fictionalism*.[1] By the way, we think that any idealization is fictional. We think of as fictional the very same infinitely many syntactical entities of a formal system.

One may wonder: if one wants to go fictionalist, why should one not prefer a fictionalism about sets rather than about *agents* and *abstracts*? Is the commitment to agents not so heavy as a commitment to sets? Our answer is that it is not.

First of all, our team of agents is countably infinite. Assuming only countably many entities we supply a categorical theory of natural numbers, while classical second order arithmetic uses a higher infinity of sets and classical first order arithmetic is non categorical.

Secondly, logical reasoning about sets presupposes the ideal possibility of referring to any of them, and, since a set is determined by its members, this possibility presupposes, in turn, the possibility of referring to all its members simultaneously. The team of agents serves just the purpose of providing a suitable way for understanding an ideal act of referring to them. On the other hand, once the team of agents has been introduced, in order to develop second order logic, the further assumption that for any plural choice the chosen objects form a set becomes superfluous. Plural choices make the job of sets without the need of any extension of the ontology. Thus the commitment to agents is much more fundamental than the commitment to arbitrary sets of individuals.

We conclude that our ontological commitment is restricted to entities of a crucial relevance for explaining a way of understanding the idealization underlying logical reasoning.

[1] On *fictionalism* see Balaguer (2008), Bueno (2009), and Yablo (2005).

Observe that the fictional flavor of our theory consists in the fact that individuals (i.e. our agents and abstracts) are treated as if they actually existed. That is essential in order to make acts of plural *simultaneous* choices performable. In contrast, acts of choice, unlike agents, are to be understood in a mere *potential* way: as said there is no realm of possible acts; an act exists only insofar as it is performable.

Again, as said, our notion of combinatorial possibility justifies immediately, without any circularity, the *comprehension principle of our language* (where A is any formula of our language):

$$\textbf{(PCP)} \quad \exists X^n \forall y_1 \ldots y_n (X^n y_1 \ldots y_n \leftrightarrow A)$$

5.10 Conclusions

On the basis of our approach to second order logic, we have here proposed a new form of *logicism* characterized by the following claims:

C.1 Finite strings of signs are entities essential for the development of logic (as emphasized by Hilbert).
C.2 Our theory of plural choices supports the claim that plural logic is pure logic.
C.3 Plural logic is adequate to structuring our closed abstracts (particular finite strings of signs) in such a way that they play the role of logical concepts of our theory **LC**.
C.4 The concepts at issue count as logical concepts, insofar as they are defined with logical means on the base of the logical relation of identity among individuals.
C.5 Natural numbers are logical entities, insofar as they are defined in terms of logical concepts.

References

Balaguer, M. (2008). Fictionalism, theft, and the story of mathematics. *Philosophia Mathematica, 17*, 1–32.

Bueno, O. (2009). Mathematical fictionalism. In *New waves in philosophy of mathematics* (pp. 59–79). Palgrave Macmillan.

Cocchiarella, N. (1985). Frege's double correlation thesis and quine's set theories NF and ML. *Journal of Philosophical Logic, 14*, 1–39.

Cocchiarella, N. (1986). Frege, russell and logicism: a logical reconstruction. In L. Haaparanta & J. Hintikka (Eds.), *Frege synthesized* (pp. 197–252). D. Reidel.

Hilbert, D. (2002). On the infinite. In J. van Heijenoort (Ed.), *From frege to goedel. A source book in mathematical logic: 1879–1931* (pp. 367–392). Harvard University Press.

Linsky, B., & Zalta, E. (2006). What is neologicism? *Bulletin of Symbolic Logic, 12*, 60–99.

Wright, C. (1983). *Frege's conception of numbers as objects*. Aberdeen University Press.

Wright, C. (1992). *Truth and objectivity*. Harvard University Press.

Yablo, S. (2005). The myth of the seven. In *Fictionalism in metaphysics* (pp. 88–115). Oxford University Press.

Chapter 6
Plural Arbitrary Reference and Mereology

Abstract This chapter argues that a certain weaker use of mereology, compared to Lewis's, supports an innocence thesis similar to plural reference. We propose a theory of *virtual mereology* (**VM**), where agents play both the role of choosers and of chosen. Using our semantics of plural choices, we interpret a formal first-order mereological language, like Goodman's *calculus of individuals*.

6.1 Introduction

In *Parts of Classes* David Lewis argues that, like logic, but unlike set theory, mereology is "ontologically innocent".

Prima facie, Lewis' innocence thesis seems to be ambiguous. On the one hand, he seems to argue that, given certain objects Xs, referring to their sum is ontologically innocent because there is no new entity as the referent of the expression "the sum of the Xs". So, talking of the sum of the Xs would simply be a different way of talking of the Xs, looking at them as a whole.

However, on the other hand, Lewis' innocence is not understood as a mere linguistic use, where sums are not reified. He himself claims that the innocence of mereology is different from that of plural reference, where reference to some objects does not require the existence of a single entity picking them out in a whole. In the case of plural quantification "we have many things, in no way do we mention one thing that is the many taken together". Instead, in the mereological case, "we have many things, we do mention one thing that is the many taken together, but this one thing is nothing different from the many" (Lewis 1991, p. 87).

But, because Lewis explicitly uses sums as objects in their own right, we think that Lewis' innocence thesis should be understood in a potential sense that, once we refer simultaneously to certain Xs we can imagine a new object representing them together. In other words, the sum of the Xs would be an ideal object generated in virtue of the very act of plural referring to the Xs. In contrast, sets are thought of simultaneously existing, independently from any act of referring. The crucial difference between the set and the sum of Xs is that the latter, unlike the former, does not determine the *xs*: different pluralities may produce the same sum.

M. Carrara, E. Martino, *Arbitrary Reference in Logic and Mathematics*, Synthese Library 490, https://doi.org/10.1007/978-3-031-66452-6_6

The aim of this chapter is to argue that—for a certain use of mereology, weaker than Lewis'—an innocence thesis similar to that of plural reference is defensible. In order to give a precise account of plural reference, we use the idea of a plural act of choice, introduced in Chap. 2. We then propose a virtual theory of mereology, where the role of individuals is played by plural act of choices of atoms. A choice is not an authentic object: its existence is merely potential and it consists in the act of performing it. Accordingly, in order to interpret a formal first-order mereological language, like Goodman calculus of individuals (*CG*), we will use our semantics of plural choices.

We argue that our development of virtual mereology, grounded on the notion of *plural act of choice*, is ontologically innocent in a way completely analogous to that of plural reference: our claim is that mereological sums—unlike atoms—are not real objects. Referring to a sum of atoms is nothing but a way of referring to certain atoms. Our approach is a constructive interpretation of first-order mereological language. It is inadequate for Lewis' mereology, because his plural quantification over all individuals is incompatible with our notion of plural choice, where only atoms are capable of being chosen.

We shall consider some attempts present in the literature to defend the ontological innocence of mereology. Such attempts suggest the possibility of a fictional ontological commitment to mereological sums, which seems to be in agreement with our approach. In fact, this is fictional insofar as it is grounded on the fictional existence of a team of infinitely many agents. We will show how to reach a version of *(CG)* that is ontologically innocent as plural quantification.

6.2 Arguments for a Fictional Ontological Commitment

We shall consider some arguments that, however certain individuals *Xs* are given, aim to deny the additional existence of the sum of the *Xs*, by introducing a weak use of sum, according to which the latter is not an individual in its own right.

The claim that the commitment to the existence of the fusion of the *Xs* is not a further commitment beyond the existence of the *Xs* is maintained in a fictional way by arguing in favor of the identification of the fusion of several individuals with the individual themselves. This identification seems to be suggested by Lewis' argument as follows:

(P1) Composition—a many-one relation—is like identity.
(P2) The commitment to sums is already presupposed in the acceptance of the objects that are summed.
(P3) Nothing could be considered more ontologically innocent than the request to accept something identical to things already accepted.
(C) Mereology is ontologically innocent.

Lewis' argument rests on the thesis (P1) of composition as identity. However, Lewis criticizes the following strong version of (P1):

(Strong Com) The predicate "are" used for the composition relation is literally the plural for the "is" of identity.

Formally:

$$\forall X \forall y((y \text{ is the sum of the } X) \rightarrow y = X)$$

One of Lewis' argument against (StrongCom) concerns the indiscernibility of identicals (InId) i.e.:

$$(\text{IdIn})\ \forall x \forall y(x = y \rightarrow \forall F(Fx \leftrightarrow Fy))$$

where the third universal quantification is of second order and "F" is a predicate variable.

> Even though – Lewis argues – the many and the one are the same portion of Reality, and the character of that portion is given once and for all whether we take it as many or take it as one, still we do not really have a generalized principle of indiscernibility of identicals... What is true of the many is not exactly what is true of the one. After all they are many while it is one (Lewis 1991, p. 87).

Consider the following example. Suppose that the number of the Xs is n, where $n > 1$. Then, the plural predicate "...are exactly n" should apply—given (InId)—to y too, but the number of y is one.

This argument shows how in Lewis' conception the sum of several individuals is a single object in its own right, but the presence of such a single object would not undermine the ontological innocence of mereology.

Lewis argues for the innocence of mereology grounding it on a weak reading of (P1):

(WeakCom) The predicate "are" used for the composition relation is *analogous* to the plural form of the "is" of identity.

The problem is that Lewis' notion of analogy is rather vague and smoky. We think that the alleged innocence of mereology is better defended within a fictional conception. An example of that might be a formulation offered by Baxter, who in Baxter (1988) argues for a way to maintain (InId) and (StrongCom). He introduces two kinds of identity, a strict and a popular identity. Baxter gives the following exemplification of the above distinction: "Suppose a man owned some land which he divides into six parcels. Overcome with enthusiasm (for the denial of composition as identity), he might try to perpetrate the following scam. He sells off the six parcels while retaining ownership of the whole. That way he gets some cash while hanging on to his land. Suppose the six buyers of the parcels argue that they jointly own the whole and the original owner now owns nothing. Their argument seems right. But it suggests that the whole was not a seventh thing" (Baxter 1988, p. 579).

A justification of (StrongCom) is to argue that to count the many strictly is to count the one loosely:

(BT) The whole is the many parts counted as one thing.

Even if Baxter argues that (**BT**) does not deny the existence of the whole, but just the additional existence of the whole, it seems to us that this popular mood does not reify the whole. Baxter's example demonstrates a weak use of the sum, not involving its existence as an entity. It seems to be a use of sums similar to that of sets in a sentence as:

(1) The set of the Germans trekking on the Plose has cardinality six hundred.

This sentence can be reformulated without involving the notion of set:

(2) The number of the Germans trekking on the Plose is six hundred.

Likewise, the sentence:

(3) I have seen a flock of seven bee eaters

can be rewritten in the form

(4) I have seen seven bee eaters.

(4) does not involve any entity denoted by the word "flock". Obviously, such examples are inadequate to defend Lewis' realistic conception of mereology. They suggest, however, the possibility of a fictional conception of mereological sums. In what follows we shall develop such a conception, which will be called virtual, grounded on the notion of arbitrary plural choice.

6.3 Arbitrary Plural Choices for a Second Order Logic

We start by making explicit a certain notion of arbitrary reference which is implicitly presupposed both by first and higher order logic. As explained in Chap. 1, first order logic implicitly presupposes the possibility of singular reference to any individual of the universe of discourse. In order to make explicit such an ideal notion of reference, we introduced an ideal agent who is supposed to be able, by means of an arbitrary act of choice, to isolate any individual.

As to second-order logic, this presupposes the possibility of referring to any set of individuals. In fact, the semantics of second-order logic quantifies over all subsets of the domain of individuals. The problem is how to refer to an arbitrary set of individuals.

Since all we know about an arbitrary set of individuals is that it is built up by its elements, reference to an arbitrary set seems to be properly realizable by simultaneously referring to its elements. The comprehension principle of second-order logic:

(CP) $\exists X \forall x (Xx \leftrightarrow A(x))$,

in the usual set-theoretical interpretation, where the second-order variables X range over all subsets of individuals, rests on the assumption that:

(**SET**) However certain individuals are given, they form a set.

The problem arises: how can arbitrary individuals be given? For the purpose we introduced in Chap. 2 an ideal team of agents, one for each individual, capable of performing an arbitrary plural choice.

For the sake of simplicity, we identify the individuals with the agents themselves. In order to perform a plural selecting choice ($p.s.c.$) each agent can choose an individual ad libitum or refrain from choose. If all agents refrain on choosing we have the empty choice (e).

The chosen individuals are said to be designated by the choice in question. Within this framework (SET) can be made explicit as follows:

(**SET***) The individuals designated by any plural choice are the members of a suitable set.

Observe that (**SET***) does not say anything about the very nature of sets. It does not claim that a set of individuals is created by a plural choice. One may maintain that the existence of sets is independent of any action and see (**SET***) as a mere explanation of how to refer to a set. However, as shown in Chap. 2, in virtue of the semantics of plural choices, one can recover classical second order logic without invoking the existence of sets. The role of sets of collecting individuals can be played by the acts of choice themselves. In the same vein, we will develop in the next section a virtual mereology **VM**, where plural acts of choice play the role of individuals.

6.4 Virtual Mereology (VM): An Informal Exposition

Imagine our team of agents. Specifically, an infinite team Σ of agents, which we shall also call mereological atoms. A plural choice consists in the simultaneous choice of an atom by each agent. The chosen atoms are said to be designated by the plural choice.

Two plural choices c_1 and c_2 are equivalent if they designate the same atoms.

As already observed, acts of choice are not individuals. Nevertheless, we can make them play the role of virtual mereological individuals: we call individuals the acts of choice and say that equivalent choices are the same individual. We treat the equivalence relation between choices as the identity relation between individuals.

Talk about individuals is to be understood as extensional talk about plural choices, i.e. talk identifying equivalent choices. If a choice designates a unique atom it is identified with the atom itself so that atoms are particular individuals. If a choice does not designate any atom (i.e. if all agents refrain to choose) it counts as the null individual (Though such individual does not occur in usual mereology, it is appropriate in our approach).

In this way, we obtain the fundamental relation of mereology. A mereological property is a property expressible logically in terms of the basic relations of $<$ and $=$.

We define the sum of c_1 and c_2, (in symbols $c_1 + c_2$) as the individual whose atoms are all those of c_1 and all those of c_2. This sum exists since there is certainly a possible choice whose designated atoms are precisely those designated by c_1 and those designated by c_2. Similarly, we define the product c_1c_2 the individual whose atoms are those common to c_1 and c_2.

More generally, there exists the sum of all individuals enjoying a mereological property P, i.e. the individual whose atoms are such that each of them is part of at least one individual enjoying P. Indeed, for any atom a it is a well-determined fact whether or not a plural choice c is possible such that a is designated by c and c enjoys P. There is therefore certainly a possible choice, that will be indicated by $\sigma x Px$, whose designated atoms are precisely those satisfying the above condition.

Similarly, there exists the product $\pi x Px$ of all objects enjoying P, i.e. the object built up from all atoms common to all objects satisfying P.

Since, as we know, acts of choice are conceived of as performable in time, we can speak of individuals of virtual mereology only in a purely potential sense. To say that there exists at least one object satisfying property P amounts to saying that a choice c satisfying P is performable.

To say that every object enjoys property P amounts to saying that, however a plural choice c is performed, it will enjoy property P. Again, we can observe that such a potential interpretation does not undermine the validity of classical logic. For, the possibility or not of performing a plural choice satisfying certain conditions is well determined by the atoms available in the domain.

One might wonder whether, and, if so, in what sense, one can speak of the cardinality of the universe of virtual individuals, and compare it with that of the universe of atoms. There is a merely potential sense in which one might claim that there are more individuals than atoms. It is possible, by means of a single plural relating choice to determine infinitely many objects. Precisely, one can produce, by a single plural relating choice (*p.r.c.*), a family of individuals indexed by atoms.

Let I the domain of atoms. If c is a plural relating choice of atoms, and i is the first component of a pair designated by c, we indicate by c_i the individual whose atoms are the ones associated with i. Thus, we get the family $\{c_i\}_{i \in I}$ of individuals c_i indexed by the atoms (or the agents).

Suppose, for the sake of simplicity, that there are denumerably many atoms and indicate them by the natural numbers. We can obtain, by means of a *p.r.c.*, a family of objects $\{c_i\}_{i \in N}$. Using the well-known method of diagonalization, one can prove the existence of an object different from all objects of the family. Precisely, let P be the property that is enjoyed by an atom i if and only if it is not an atom of c_i. It is certainly possible to perform a choice c whose designated atoms are exactly those enjoying P. Such a choice is an individual different from all c_i.

So, we can say that the individuals constitute a non-denumerable infinity in the sense that, however a countable family of individuals is determined by means of a plural relating choice, it is always possible to determine an individual different from all components of that family. It follows that no plural choice can give simultaneous existence to all individuals.

Since our team of agents is countable, so is any plurality of individuals generated, as above, by a *p.r.c.* Nor can we extend the notion of plural choice of atoms to that of plural choice of individuals by introducing an agent for every individual. In fact, as we know, a plural simultaneous choice requires the simultaneous existence of all agents, while our individuals have only a potential existence.

6.5 The Calculus of Individuals with Atoms (*CG*)

Our virtual mereology can be seen as a constructive version of the axiomatic theory of mereology developed by N. Goodman in Goodman (1977) called "calculus of individuals" (*CG*) where the overlap relation is assumed as primitive.

In the present section we consider a version of (*CG*) with atoms but without sets, where the part relation (<) is primitive, while the other mereological notions and identity (=) are defined in terms of the part relation.

Precisely, we introduce the following definitions:

Def.1 $xoy =_{df.} \exists z(z \neq e \wedge z < x \wedge z < y)$ (x overlaps y)
Def.2 $x = y =_{df.} x < y \wedge y < x$ (x is identical to y)
Def.3 At (x) $=_{df.} x \neq e \wedge \forall y(y < x \rightarrow ((y = x) \vee y = e))$ (x is an atom)

Two non-overlapping individuals are said to be mutually discrete:

Def.4 $x \wr y =_{df.} \neg(xoy)$ (x is discrete from y);
Def.5 $x << y =_{df.} (x < y \wedge \neg(y < x))$ (x is proper part of y)

Sum or fusion has been defined in Sect. 6.3.

Axioms:

A1 Any set of logical axioms for first-order predicate calculus without identity.
A2 $x < x$
A3 $x < y \wedge y < z \rightarrow x < z$
A4 $\forall z(Atz \wedge z < x \rightarrow z < y) \rightarrow x < y$
A5 $\forall z(zox \leftrightarrow zoy) \rightarrow (A \rightarrow A[y/x])$

where $A[y/x]$ is any formula obtained from A by replacing some occurrence of x free in A with an occurrence of y.

A6 For every formula $F(x)$, there is one and only one individual which is the sum of all individuals satisfying $F(x)$.

Such an individual will be denoted by $\sigma x F(x)$. In particular, taking for $F(x)$ the formula:

$$(x = y \vee x = z)$$

we get the sum of y and z which will be simply indicated by $y+z$. Finally, it is easily seen that, if the F share a common part, there exists the product of the individuals satisfying $F(x)$, i.e. the individual built up from all atoms shared by them. We shall indicate this product by $\pi x F(x)$; in case of two individuals y and z simply by yz. It follows from (CGA4) that:

(Theorem 1) $\forall x(x \neq e \rightarrow \exists y(Aty \wedge y < x))$ (each individual has an atomic part)

Proof Suppose, for the purpose of reductio, that x is an individual without atoms. It follows from (A4) that x is part of every y, so x has no proper parts. Therefore, it is itself an atom, which is absurd.

It is clear from Sect. 6.3 that all axioms are satisfied by **VM**.

6.6 Conclusions

In the present chapter we have tried to take at face value Lewis' claim that mereological sums are identifiable with the objects composing them. To this end, we have developed a virtual mereology (**VM**), where atoms are the sole genuine objects, while composed objects are nothing but pluralities of atoms. So, the innocence of (**VM**) is inherited from the innocence of the plural interpretation of second-order logic, where second-order variables range over individuals plurally.

Our virtual mereology is inadequate to interpret Lewis' use of mereology, since this essentially exploits the existence of sums as genuine individuals. In fact, Lewis tries to defend the innocence of mereology by arguing that, though sums are genuine objects, nevertheless the commitment to their existence is already implicit in the commitment to the existence of atoms. However, we will show in Chap. 8 how Lewis' alleged innocence is hardly defendable.

Our version of mereology, as well as Lewis' mereology, essentially exploits the atomic assumption, according to which every individual is built up from atoms. Observe, however, that such an assumption, as it is used in the present chapter, does not concern the nature of the atoms. Any genuine individuals can play the role of atoms (in particular our agents), insofar as we are not interested in their internal structure, but in the mereological individuals built up from all possible fusions of them. However certain *genuine* individuals are given, they can count as atoms and their fusions as *virtual* individuals.

References

Baxter, D. L. (1988). Identity in the loose and popular sense. *Mind, 97*, 575–582.
Goodman, N. (1977). *The structure of appearance*, vol. 53. Springer.
Lewis, D. (1991). *Parts of classes*. Blackwell.

Chapter 7
Grounding Megethology on Plural Arbitrary Reference

Abstract D.K. Lewis, in his "Mathematics is Megethology," combines mereology with plural quantification to reconstruct set theory, creating a *megethology* with enough expressive power to explore hypotheses about the size of reality. This chapter presents a new approach to megethology based on the theory of plural arbitrary reference developed earlier in the book. Our approach demonstrates how megethology can be founded on plural arbitrary reference without relying on mereology.

7.1 Introduction

Lewis reconstructs set theory in "Mathematics is megethology" (Lewis 1993). By combining mereology with plural quantification, he introduces a powerful framework *megethology,* "so called because it turns out to have enough expressive power to let us express interesting hypotheses about the size of Reality" (Lewis 1993, p. 3). Within this framework, Lewis develops a structuralist class theory, in which the role of classes is played by individuals. Thus, accepting Boolos' thesis that second-order logic, interpreted in terms of plural quantification, does not involve the notion of class, Lewis achieves an ontological reduction of classes to individuals.

Lewis maintains that both mereology and plural quantification are part of logic, but this claim is highly controversial and has been criticized by several authors.[1] In particular, we will show that the combined use of mereology and plural quantification assures that if the domain of individuals is infinite then it is uncountable. It seems to be implausible that such a result holds in virtue of pure logic. The general notion of individual is somewhat vague and inadequate to justify the main axiom of mereology, according to which any plurality of individuals, however given and heterogeneous, form a further individual, their sum. Furthermore, the crucial

[1] See Parsons (1990) and Resnik (1988) for a criticism to second order logic, and Carrara and Martino (2011) for a criticism to the thesis that mereology is part of logic.

M. Carrara, E. Martino, *Arbitrary Reference in Logic and Mathematics*,
Synthese Library 490, https://doi.org/10.1007/978-3-031-66452-6_7

distinction between pluralities and classes has not been adequately elaborated by Boolos and Lewis to justify their claim that plural quantification is ontologically innocent. They failed to dispel the suspicion that pluralities are nothing but masked classes.

Lewis' ground of megethology, consisting of both mereology and plural quantification, turns out to be overloaded. In this chapter we first introduce Lewis' megethology, then we show how megethology is developable using plural quantification without mereology. In the next chapter we will show how megethology is developable using mereology without plural quantification.

We will exploit our approach to plural quantification based on our theory of semantics of acts of choices (**SAC**). In particular, we will use our notion of *plural selecting choice (p.s.c.)* and of *plural relating choice (p.r.c.)* developed in Chap. 2.

7.2 Lewis' Megethology in a Nutshell

As said before, by combining mereology with plural quantification, Lewis introduces megethology, a powerful framework in which one can formulate strong assumptions about the size of the universe of individuals (corresponding to the existence of strongly inaccessible cardinals). Within this framework, Lewis develops a structuralist class theory, in which the role of classes is played by certain fusions of atoms and the membership relation is defined in mereological terms.

Mereology with plural quantification (MPQ) can be formalized into a first order language with identity with two sorts of variables:

Singular variables: x, y...
Plural variables: X, Y...

We introduce the primitive non-logical constants:

$\leqslant$: $x \leqslant y$ is to be read "x is part of y";
η : $x \eta X$ is to be read "x is one of Xs".

We define:

$x<y$ (x is a proper part of y) if$_{df}$ $(x \leqslant y \wedge x \neq y)$;
x is an atom if$_{df}$ it has no proper parts;
$X \sqsubseteq Y$ (the Xs are among the Ys) if$_{df}$ $\forall x(x\eta X \rightarrow x\eta Y)$;
$X \sqsubset Y$ (the Xs are among, but not all, the Ys) if$_{df}$ $(X \sqsubseteq Y \wedge X \neq Y)$;
$X \sqcup Y$ (the sum of the Xs and the Ys) is the plurality Z such that $\forall x(x\eta Z \leftrightarrow x\eta X \vee x\eta Y)$;
$x \circ y$ (x overlaps y) if$_{df}$ $\exists z(z \leqslant x \wedge z \leqslant y)$
y is a sum of the Xs, in symbols σX, if$_{df}$ each of the Xs is a part of y and each part of y overlaps one of the Xs. Formally:

$$\forall x(x\eta X \rightarrow x \leqslant y) \wedge \forall x(x \leqslant y \rightarrow \exists z(z\eta X \wedge x \circ z)$$

πX (the product of the Xs) is the fusion of all common parts of all the Xs, provided there is at least one part common to all the Xs.

We will first introduce the axioms for mereology:

Axiom 1 $\leqslant$ is a partial order (*reflexive, antisymmetric, transitive*).
Axiom 2 *(Atomicity)* Every individual has an atomic part.
Axiom 3 For every X there is a unique sum of the Xs.

From axioms 2 and 3 it straightforwardly follows that every individual is the fusion of its atoms.

Besides, the crucial *Comprehension* for pluralities:

A4(PCA) Given any formula ϕ of our language, satisfied by at least an individual, there is the plurality of all individuals satisfying it:

$$\exists x\phi \rightarrow \exists X\forall x(x\eta X \leftrightarrow \phi)$$

Finally, we list Lewis' definitions and axioms concerning the size of the universe.

Definitions

An individual is *small* if$_{df}$ there is no bijection between its atoms and all atoms of the universe V;
A plurality X is *small* if$_{df}$ there is no bijection between the Xs and all atoms of the universe V;
An individual x is *infinite* if$_{df}$ there is a bijection between its atoms and the atoms of a proper part of x.

Axioms

Axiom 5 The fusion of few small things is small.
Axiom 6 The parts of a small thing are few.
Axiom 7 Something small is infinite.

7.3 Megethology with Choices

Let us introduce, using pluralities, the basic notions and axioms of megethology. Let S be the universe of individuals.

A plurality X is *small* if$_{df}$ there is no plural injection of S to X. Otherwise it is *large*.
A plurality X is infinite if$_{df}$ there is an injection of X into a proper subplurality of X.

By means of a binary *p.r.c.* a single individual can represent a plurality of individuals, precisely the plurality of its correlates. In this way we want to elaborate Black's idea of a plural reference understood as a "plural pointing".

Call "representation" a binary *p.r.c.* such that no two distinct individuals represent the same plurality.

Question: could some possible representation satisfy the condition that, however a *p.s.c.* is performed, the chosen individuals have a representative?

The answer is no, as we will see.

However, we will show that, if **ZFC** is consistent, one can consistently assume the existence of a representation such that all small (in a sense to be specified) pluralities are representable. Relative to such a representation our agents form a secondorder model of **ZFC**.

We assume that a representation D is possible, such that the axioms below are satisfied. From now on the reference to such a D will be understood. The following axioms assure that the size of the universe S of individuals is suitably large for reconstructing set theory:

Axiom 8 Every small plurality has a representative.

Axiom 9 Some infinite plurality is small.

Axiom 10 The plurality of the representatives of all subpluralities of a small plurality is small.

Axiom 11 Let X be a small plurality of representatives of small pluralities. The plurality Y, such that for all y, $y \eta Y$ iff y is represented by one of the Xs, is small.

Lemma 7.1 *Some pluralities have no representatives.*

Proof Suppose—by way of contradiction—that every plurality has a representative. If so, we can reproduce the Russell paradox by taking the plurality X of all representatives that do not represent themselves and wondering whether the representative of the Xs is or not one of the Xs.

Corollary 7.1 *Given any large plurality* X*, some subpluralities of* X *have no representatives*

Proof Let X be large. There is an injection f of the plurality S of all individuals into X. Suppose, by reduction, that every subplurality of X has a representative. We can define a new representation D' by taking as the representative of any plurality Y the D-representative of the corresponding subplurality $f(Y)$ of X. So every plurality has a D'-representative, against *lemma* 7.1.

Theorem 7.1 *Let* X *be a small plurality of representatives of small pluralities. The plurality* Y*, such that for all* y*,* $y \eta Y$ iff y *is represented by one of the* Xs*, is small.*

Proof Suppose, by way of contradiction, that Y is large. We will show that it is possible to define a new representation D', such that every subplurality of Y has a D'-representative, against *corollary* 5. Let V be any subplurality of Y. V meets any plurality represented by some x of the Xs into a small plurality represented by some x_V. The plurality Z of all x_V, when x ranges over all the Xs, is small. Let z be the representative of the Zs. As V is determined by Z, we can define D' by taking z as the D'-representative of V. Thus any subplurality of Y has a D'-representative.

7.4 An Interpretation of *ZFC* Set Theory

We define the plurality U of sets as the individuals belonging to every plurality X satisfying the following conditions:

(i) $o\eta X$ where o is the representative of the empty plurality,
(ii) if x is the representative of a small subplurality of X, then $x\eta X$.

It follows that every set is the representative of a small plurality of sets.
We use $\alpha, \beta, \gamma\ldots$ as variables for sets.
Define *membership* as the inverse of representation D:

$$\alpha \in \beta \text{ if}_{df}\ \alpha \text{ is one of the individuals represented by } \beta.$$

So o is the empty set and will be indicated by the usual symbol $\varnothing$. Let us verify the axioms of *ZFC*. From axioms 1 and 3 it follows immediately the

Axiom 12 of Extensionality Two sets are identical iff they have the same members.

Since two sets form a small plurality, the following axiom follows:

Axiom 13 of Pair For all α, β there is a set $\{\alpha, \beta\}$ whose members are exactly α and β.
Axiom 14 of Union For every nonempty set α, there is a set β whose members are all members of members of α.

Proof From *theorem* 7.1 it follows that the plurality X of members of members of α is small, so it forms a set.
From *axiom* 11 it follows immediately that:

Axiom 14 of Powerset For every set α there is a set β whose members are the subsets of α. Any subplurality of the members of a set is small, so it forms a set.

In particular:

Axiom 15 of Separation The members of a set satisfying any formula form a set.

Since any plural function maps small pluralities into small pluralities, we get in particular:

Axiom 16 of Replacement If ϕ is a functional formula, i.e., such that $\forall x \exists! y \phi$, the ϕ-image of a set is a set.

From *axiom* 8 it follows directly the *axiom of choice*:

Axiom 17 of Choice If α is a set of nonempty pairwise disjoint sets, there is a set β that shares a unique member with every member of α.

Proof A plural choice function (see *axiom* 7) maps the members of α into a small plurality of sets, which are the members of the required set β.

Define the plurality $\mathbb{N}$ of Zermelo natural numbers as the least plurality X of sets satisfying the following clauses:

(i) $\varnothing \in X$;
(ii) if $x \in X$ then $\{x\} \in X$.

One can verify that $\mathbb{N}$ is Dedekind infinite according to the plural function of ordered pairs $(x, \{x\})$. One can also verify that there is an embedding of $\mathbb{N}$ into any infinite plurality, in particular (by *axiom* 9) into an infinite small plurality. So the plurality $\mathbb{N}$ is small and therefore determines a set.

Consequently, we get the axiom:

Axiom 18 of Infinite Some set is infinite.
Axiom 19 of Regularity If $\alpha \neq \varnothing$, there is a $\beta \in \alpha$, such that $\beta \cap \alpha = \varnothing$.

Proof The plurality of $\varnothing$ and the sets α, for which there is a $\beta \in \alpha$, such that $\beta \cap \alpha = \varnothing$, satisfies clauses (i), (ii) of the definition of U. Thus $X = U$.

7.5 Categoricity

Given the universe S of individuals, the universe U of sets depends on the representation relation D. But one can easily recognize that if D^* is a new representation satisfying the same axioms for D, the universe U^* relative to D^* is isomorphic to U. In fact, let R be the least plural D-relation such that

(i) $R(o, o^*)$;
(ii) if H is a small subplurality of R, x the D-representative of the first components, x^* the D^*-representative of the second components of the ordered pairs of H, then $R(x, x^*)$.

It is easily seen that R is an isomorphism.

7.6 Concluding Remarks

Our approach to megethology starts from the observation (developed in Chap. 2) that the possibility, in principle, of referring to any individual of the universe of discourse of a mathematical theory is essentially, although implicitly, presupposed in mathematical reasoning. In order to make such presupposition explicit and to clarify the sense of the locution "in principle," we have taken as individuals a team of ideal agents. By means of this device, we have obtained an idealized version of

Kripke's notion of direct reference based on the crucial distinction between acts and things.

Our framework provides a new perspective for understanding the ontological innocence of plural reference and plural quantification. Speaking of a *plurality* of individuals is nothing but a linguistic device for referring to the individuals selected by an act of choice. Moreover, our team of agents provides an intuitive means for simulating relations, without introducing the latter as a particular sort of entities.

One could perhaps object that our approach to pluralities is far from being ontological innocent just because of our commitment to a so large team of infinitely many agents. Of course we cannot deny that agents are entities. In this respect our approach is not in agreement with Lewis' claim that the transition from singular to plural reference is free of any further ontological commitment: such a transition requires a surplus of idealization. However, it would be a mistake to conclude that our commitment to agents is as strong as the traditional commitment to classes.

First, as observed in Chap. 2, since all that we know about an arbitrary class of individuals, set-theorethically understood, is that it is built up by its members, singular reference to a class must be obtained through plural reference to its members; and our team of agents serves just the purpose of making the idealization of plural reference explicit.

Second, while classes are abstract entities, agents are concrete, though imaginary, entities. Our approach shows how, adopting a suitable structuralist view of mathematics, one can avoid the assumption of abstract entities: infinitely many concrete entities are sufficient for modeling any mathematical theory, provided that their infinity is suitably large.

References

Carrara, M., & Martino, E. (2011). On the infinite in mereology with pural quantification. *The Review of Symbolic Logic, 4*, 54–62.

Lewis, D. (1993). Mathematics is megethology. *Philosophia Mathematica, 1*(1), 3–23.

Parsons, C. (1990). The structuralist view of mathematical objects. *Synthese, 84*(3), 303–346.

Resnik, M.D. (1988). Second-order logic still wild. *The Journal of Philosophy, 85*(2), 75–87.

Chapter 8
The Mereological Foundation of Megethology

Abstract In this chapter we show how, assuming the existence of a pairing function on atoms, as the unique assumption non expressed in a mereological language, a mereological foundation of set theory is achievable within first order logic. Furthermore, we show how a mereological codification of ordered pairs is achievable with a very restricted use of the notion of *plurality*. Finally, in the last section of this chapter we show that, adopting a relativistic notion of atom, according to which any individual can play the role of atom, one can reconstruct Cantor's theorem and prove that, given any infinite, mereology with plural quantification (MPQ) guarantees the existence of large and large infinities. This result seems in conflict with the alleged ontological innocence of MPQ.

8.1 Introduction

Though in the Lewis' work on megethology plural quantification is used extensively, a careful analysis shows that it plays an essential role only as a device for providing a mereological codification of ordered pairs. This is surely an interesting step. However, the notion of ordered pair is certainly not one of the most problematic notions of set theory. Therefore giving up plural quantification and taking ordered pairs as primitive seems to be a natural move in order to make the mereological conception of set theory more perspicuous.[1]

Accordingly, in this chapter we will propose a version of megethology within first order logic with identity by using, besides the mereological part-whole relation $\leqslant$, a primitive *pairing function* (,) from ordered pairs of atoms to atoms. In this framework we will develop a structuralist interpretation of set theory with proper classes (in the Von Neumann-Goedel-Bernays style).

[1] Hellman (1989) and Feferman and Hellman (1995) also incorporated primitive pairing in their systems.

M. Carrara, E. Martino, *Arbitrary Reference in Logic and Mathematics*, Synthese Library 490, https://doi.org/10.1007/978-3-031-66452-6_8

In the last section we will show how a suitable mereological codification of ordered pairs of atoms by atoms is achievable with a restricted use of the notion of plurality that does not involve plural quantification.

8.2 Mereological Axioms

We re-introduce the language of megethology, this time with a pairing function (,) from ordered pairs of atoms to atoms. The language $\mathcal{ML}$ of megethology is a first order language with identity. Variables *x*, *y*,... range over individuals. The primitive non-logical constants are the binary relation $\leqslant$ ($x \leqslant y$ is to be read "*x* is part of *y*") and the pairing function (,) from ordered pairs of atoms to atoms.

Axiom 1 $\leqslant$ *is a partial order* (reflexive, antisymmetric, transitive).

$x < y$ (*x* is a proper part of *y*) if$_{df.}$ $x \leqslant y$ and $x \neq y$.
x is an atom if$_{df.}$ it has no proper parts.

Axiom 2 *Every individual has an atomic part.*

Fusion schema:
Axiom 3 $\exists z\phi \rightarrow \exists! y \forall x(x\ atom \rightarrow (x \leqslant y \leftrightarrow \exists z(\phi \wedge x \leqslant z))$, where ϕ is a formula with possible parameters and without *x*, *y* free.

Such *y* is the *fusion* of the ϕ's. (in words: if there are some ϕ's, there is the fusion of all ϕ's).

From 2 and 3 it follows that every individual is the fusion of its atoms.
Besides it follows that:

Axiom 4 Extensionality Two individuals are equal iff they have the same atoms.

Axiom 5 Complementation If x is a proper part of y the complement of x in y is the fusion of all atoms of y that are not part of x

If all ϕ's have some part in common, we define the *product* of the ϕ's as the fusion of all their common parts.

Axiom for the Pairing Function $(x, y) = (u, z) \leftrightarrow x = u \wedge y = z$, where *x*, *y*, *u*, *z* are atoms as well as the codes of the ordered pairs.

8.3 Axioms on the Size of the Universe

Using the pairing function, we can correlate atoms by means of individuals. An *atom-relation* (briefly *a-relation*) is a fusion of ordered pairs of atoms.

In particular, a function is a *many-one-a-relation*. Such a function can be extended in a natural way to non atomic individuals by mapping an individual into the fusion of the images of its atoms.

Following Lewis (1991) we say that an individual is *small* if there is no *1-1-a-relation* between its atoms and all atoms of the universe V.

An individual x is (Dedekind) infinite if there is a *1-1-a-relation* between its atoms and the atoms of a proper part of x.

The following axioms assure that V (the fusion of all atoms) is very large:

Axiom 6 Some small individual is infinite.

Let r be an *a-relation*. We say that the atom x *r-represents* the individual y if, for all atoms z, $z \leqslant y$ *iff* $(z, x) \leqslant r$.

The following axiom of representability enables us to represent small individuals by atoms:

Axiom 7 There is an *a-relation r such that every small individual has a unique r-representative.*

Such an *a-relation* will be called a *representing a-relation.*

Axiom 8 If x is small and r is a representing a-relation, then the fusion of all r-representatives of parts of x is small.

It is worth noting that Axioms 7 and 8 are independent of the preceding ones, as the following counter-examples show:

About 7. Suppose that the size of the universe is a cardinal α, with $\aleph_0 < \alpha < \mathfrak{c}$ (the cardinality of the continuum), what is compatible with the failure of the *continuum hypothesis*. Then 6 holds because any countable individual is small. But, in order to represent by atoms the parts of a denumerable individual, $\mathfrak{c}$-many atoms are needed; so 7 fails.

About 8. Consider set theory without the power axiom. This theory can be interpreted in a universe of sets, where all infinite sets are denumerable. Interpreting small individuals as sets and atoms as singletons, 6 and 7 hold, but 8 fails, because the singletons of the parts of a denumerable set do not form a set.

Axiom of Choice 9 There is an a-function that maps each representative atom into an atom of the represented individual.

8.4 An Interpretation of Set Theory

Let r be a representing *a-relation*. Let u be an atom that is not an *r-representative* (since V is infinite, we can assume that such an atom exists). Relative to r and u we define a model of set theory. We shall refer to the Kelley-Morse axiomatization of class theory, as formulated in Kelley and Stone (1955).

Let U be the mereological product of all individuals x such that (i) $u \leqslant x$ and (ii) the *r-representative* of any small part of x is a part of x.

In other words, U is the smallest individual, including u, and closed under *r*-representation.

Take as *classes* the parts of U and use capital letters $X, Y, Z, \ldots$ as variables for classes. Define *membership* as follows:

$X \in Y$ if$_{df.}$ X is small and its *r-representative* is part of Y.

A class X is a *set* if$_{df.}$, for some Y, $X \in Y$.

It follows immediately that a class is a set *iff* it is small.

We will prove that all the main axioms of the Kelley-Morse class theory hold.

Lemma 8.1 *Every atom of U, different from u, is an r-representative.*

Proof Let y be an atom, different from u, which is not an r-representative. The fusion x of all atoms of U, different from y, satisfies conditions (i), (ii) of the definition of U. Hence y is not part of U.

So u is the unique class without members and will be denoted by 0.

Axiom 10 If two classes have the same members, they are identical.

Proof It follows from the lemma that classes with the same members are fusions of the same atoms.

Axiom 11 Let ϕ be a formula of class theory, whose free variables are among X, $Z_1, \ldots, Z_n$. For all classes $Z_1, \ldots, Z_n$, there is a class Y whose members are just the sets X such that ϕ.

We introduce the usual notation $Y = \{X: \phi\}$.

Proof Given $Z_1, \ldots, Z_n$, if there is some set X such that ϕ, then Y is the fusion of the *r-representatives* of all sets satisfying ϕ; otherwise $Y = 0$.

Standard class-theoretical notions are defined as usual.

$X \subset Y$ means that X is a subclass of Y, including the case that $X = Y$.

$2^X =_{df.} \{Y : Y \subset X\}$.

From axiom 8 we get immediately

Axiom 12 If X is a set, so is 2^X.
$\{X, Y\} =_{df.} \{Z : Z = X \vee Z = Y\}$.
Axiom 13 of pairs $\{X, Y\}$ is a set.

Proof $\{X, Y\}$ is 0, if both X and Y are proper classes. Otherwise it is the fusion of at most two atoms, so it is small.

Ordered pairs are defined in the usual way. Relations are (possibly proper) classes of ordered pairs. Functions are many-one relations. If the domain of a function is small, so is clearly its range. Thus

Axiom 14 If the domain of a function is a set, so is its range.

Axiom 15 If X is a set, so is $\bigcup X$.
Proof Suppose, by reduction, that $\bigcup X$ is a proper class. Then it has the same size as U and we can suppose $U = \bigcup X$. For any class Y, the intersections of Y with the members of X are (by 14) the members of a set $Y*$. By mapping Y into $Y*$, we get a bijection between classes and sets. This leads to an absurdity, arguing as in Cantor's theorem.

Axiom 16 If $X \neq 0$, there is a member Y of X such that $X \cap Y = 0$.

Proof Let x be the fusion of all atoms n satisfying the following condition: for every individual y such that $n \leqslant y$, there is an atom $m \leqslant y$ such that, for no atom $p \leqslant y$, $(p, m) \leqslant r$. It is easily seen that x satisfies conditions (i), (ii) of the definition of U. So $U \leqslant x$ and the theorem follows.

By using the foregoing set-theoretical axioms, one can define ordinal numbers and show that they are the members of a proper class, which is well-ordered by membership. Since proper classes have the same size as U, there is a well-order of U and the axiom of choice follows:

Axiom of Choice 17 There is a choice function that assigns to every set one of its members.

8.5 Categoricity

Our model U of class theory is relative to the arbitrary choice of r and u. We want to show that the models obtained by changing r and u are isomorphic.

Categoricity Let r', u' be an a-relation and an atom satisfying the same conditions as r, u and let U' be the relative model of class theory. Then there is an a-function h (extended to individuals in the natural way) through which U and U' are isomorphic.
Proof We will define, by transfinite induction, for every ordinal α of U, a $1-1-$ a-function h_α from a part of U to a part of U' such that

1. $h_\alpha(u) = u'$;
2. If $h_\alpha(a) = a'$ where a, a' are atoms, then either $a = u$ and $a' = u'$, or a is the r-representative of x, a' the r'-representative of x' and $h_\alpha(x) = x'$.

Put:

1. $h_0(u) = u'$;
2. $h_{\alpha+1}$ extends h_α as follows: for every x, x' such that $h_\alpha(x) = x'$, put $h_{\alpha+1}(a) = a'$, where a is the r-representative of x and a' the r'-representative of x'.
3. $h_\alpha = \cup_{\beta<\alpha} h_\beta$, for α limit.

One recognizes, by transfinite induction, that all functions are 1-1 and satisfy clauses (i) and (ii).

We claim that the fusion h of all h_α is the required isomorphism.

Proof Let D be the domain of h and, for each α, D_α the domain of h_α. Of course, $D \leqslant U$. In order to show that $D = U$, we have to prove that D is closed under r-representation. Observe that all D_α are small. For, if D_α is small, so is $D_{\alpha+1}$, by 2.3; besides, the fusion of a bounded sequence of small individuals is small. Hence, by induction, every D_α is small. Now, let x be a small part of D. Since the atoms of x are few and each of them occurs in some D_α, x is part of some D_α and its representative is part of $D_{\alpha+1}$. Thus D is closed under r-representation; and hence $D = U$.

Similarly, the codomain of h is the whole U'. For, observe that the height of the ordinals of U' is the same as that of the ordinals of U, since it is determined by the size of the universe of atoms.

Besides, it is plain that h preserves membership, so that it is an isomorphism between U and U'.

It is worth noticing that our result of categoricity presupposes an absolute sense of the mereological notions of *part* and of *fusion*. In virtue of that non-standard models are ruled out.

Besides, notice that categoricity shows the independence of the set theoretic structure of the choice of the representing relation and the non-representing atom, but the structure still depends on the size of the universe of atoms. Changing the universe of atoms one gets a result of *quasi-categoricity*: one of the models is isomorphic to a segment of the other.[2]

Thus, the mereological structure of individuals determines, up to isomorphism, a model of class theory.

We can conclude that the conception of a big universe of individuals, mereologically structured, provides, with the help of any given pairing function, an adequate framework for a structuralist foundation of class theory.

8.6 Consistency of Megethology

In this short section we want to observe that megethology is, in turn, interpretable in Kelley-Morse class theory, so that it is consistent relatively to the latter.

For, take non-empty classes as individuals, singletons of sets as atoms, subclasses as parts. Define the pairing function by putting

$(x, y) =_{df.} \{\{x\}, \{x, y\}\}$, where x, y are singletons.

As it is known, it is provable from Kelley-Morse axioms that a class is proper iff it is equinumerous with the total class U (see Fraenkel et al., 1973, p.137). So sets

[2] The situation is very similar to Zermelo's *quasi-categoricity*. See on that Hellman (2002).

are small classes and we can define a representing a-relation r as follows:

$(x, y) \leqslant r\ iff$, for some set z, x is the singleton of a member of z and y the singleton of z.

All axioms of megethology are straightforwardly verifiable.

8.7 A Mereological Codification of Ordered Pairs

Our development of megethology, without plural quantification, with the help of a primitive pairing function, puts forward that Lewis' use of plural quantification serves only the purpose of supplying a mereological codification of ordered pairs. We want to show that even for this purpose plural quantification is overwhelming.

Lewis codifies ordered pairs of atoms as biatoms, relative to certain given pluralities of individuals. In this way he can identify a binary relation with a certain plurality of biatoms. So quantification over relations is understood as plural quantification. We will show that, with a suitable modification of Lewis' strategy, ordered pairs of atoms can be codified by single atoms (instead of biatoms). In this way an a-relation can be understood as a fusion of atoms, i.e. as a single individual, and quantification over relations is reduced to first order quantification.

Following a Burgess' idea (Burgess et al., 1991), assume that the universe U of atoms is divided in three microcosms U_1, U_2, U_3 of the same cardinality of U and pairwise non-overlapping.

Let f be a 1-1 map from U_3 to U_1 and g a 1-1-map from U_3 to U_2, understood as pluralities of biatoms. So every atom of U_3 has a counterpart in U_1 and a counterpart in U_2. Let A be a plurality of three-atoms satisfying the following conditions:

(i) for every x in U_1, y in U_2 there is a unique z in U_3 such that $x + y + z$ is in A;
(ii) every atom in U_3 can occur in a unique three-atom of A.

Relative to the pluralities f, g and A, we can codify the ordered pairs of U_3: if x and y are atoms of U_3, take as ordered pair (x, y) the unique z in U_3 such that $f(x) + g(y) + z$ is in A. In this way the ordered pairs of atoms of U_3 are defined mereologically, relative to a finite number of fixed pluralities, i.e. f, g, and A.

Of course, the restriction of ordered pairs to U_3 is sufficient for the reconstruction of megethology: since the size of U_3 is the same as that of U, once ordered pairs are defined, one can build megethology on the universe U_3.

8.8 On the Infinite in Lewis' Mereology

In this section, we want to show how Lewis' combination of mereology and plural quantification can guarantee that, given any infinity of individuals, the existence of large and large pluralities of individuals follows.

A mereological individual is conceived of as composed by certain atoms. These are characterized as individuals without proper parts. However, mereology does not exploit the real nature of atoms, but rather their behavior with respect to the relation of part. The notion of part is introduced, in turn, as any partial order relation $\leqslant$ satisfying certain axioms. For an individual a atomicity is characterized by the condition:

$$\forall x(x \leqslant a \rightarrow x = a).$$

It seems therefore that atomicity can be understood, mereologically, as the mere role played by an individual with respect to a certain order relation. In other words, any individual can be regarded as an atom by prescinding on its internal structure. We implicitly adopted such point of view, call it *relativistic mereological perspective*, in our virtual mereology **VM** (see on that Chap. 6), when we took as atoms our imaginary agents, obviously disregarding their human composition.

We will exploit this relativistic perspective for a mereological reconstruction of Cantor's theorem.

We will use the symbols introduced in Chap. 7:

Singular variables: x, y...
Plural variables: X, Y...

We introduce the primitive non-logical constants:

$\leqslant$: $x \leqslant y$ is to be read "x is part of y";
η : $x\eta X$ is to be read "x is one of Xs".

Let A be an infinite plurality of individuals, which we will call atoms. MPQ guaranties the existence of a universe U of pluralities satisfying the following axioms:

Axiom 18 $\leqslant$ is a partial order (*reflexive, antisymmetric, transitive*);
Axiom 19 For every atom x, $(y \leqslant x \rightarrow y = x)$. (An atom has no proper parts);
Axiom 20 $(x \leqslant y)$ iff, for all $z\eta A$, $(z \leqslant x \rightarrow z \leqslant y)$;
Axiom 21 for every y there is $x\eta A$ such that $x \leqslant y$ (each individual has an atomic part);
Axiom 22 For every plurality $P \subseteq A$ of atoms there is an x such that, for all $y\eta A$, $y \leqslant x$ iff $y\eta P$. x is the sum of the Ps: $x = \sigma Ps$;

Assume, for the moment, the notion of function as primitive. Later we will propose a mereological definition of function suggested by the present relativistic perspective.

In the following we prove a mereological version of Cantor's theorem.

(CT) Let A and U be as above. The size of U is larger than the size of A.

Proof Assume, by way of contradiction, the existence of a bijective function $f : A \to U$. Suppose, by reduction, that, for all $x\eta A$, $x \leqslant f(x)$. Since f is bijective, for every $y\eta A$ there is $z\eta A$ such that $f(z) = y$. So $z \leqslant y$ and, by (A. 2), $y = z$. So, for all $x\eta A$, $f(x) = x$, against the surjectivity of f onto U (Not all Us are atoms!). Therefore, for some $x\eta A$, $\neg x \leqslant f(x)$ and so there is the plurality K of all atoms x such that $\neg x \leqslant f(x)$. Let k be the sum of the Ks and let $h\eta A$ be such that $f(h) = k$. The contradiction $h\eta K \leftrightarrow \neg h\eta K$ follows.

Since all Us can, in turn, play the role of atoms with respect to a new part-relation, the above argument can be reiterated over and over to prove the existence of larger and larger infinities of individuals.

In particular, given a countable infinity of individuals, Lewis'mereology with plural quantification guarantees the existence of an uncountable infinity of objects.

However, the history of the philosophy of mathematics seems to show that the claim that any infinite is countable can be held within a respectable conception of mathematics. Such famous advocates as Kronecker, Poincaré, Skolem, among others, have defended this claim.

Besides, an uncountable infinite allows to reformulate in MPQ the continuous hypothesis CH, whose possible truth is a question that seems to transcend the frame of pure logic. But, since according to Lewis, all pluralities are thought of as simoultaneously existent, the acceptance of an actual countable infinite forces the acceptance of an actual uncountable infinite as well.

In contrast, this criticism does not affect our virtual mereology **VM**. In fact, as we saw in Chap. 6, there is in **VM** a crucial difference between authentic individuals and virtual mereological individuals. Authentic individuals are our agents, thought of as endowed of an autonomous existence. Virtual mereological individuals, on the other hand, are nothing but acts of plural choices. These do not create any new authentic individuals. Regarding such acts as virtual individual with respect to extensional equality is nothing but a *façon de parler*.

As to a plurality of objects, this is determined, as we saw, by a simultaneous choice of ordered pairs of atoms. So, if the plurality of atoms is countable, so is any plurality of objects. In this sense **VM** allows to recover the ontological innocence of mereology with plural quantification.

It may be of some interest to show how the above relativistic perspective may be exploited for defining a mereological notion of ordered pair and hence of function.

To the purpose, let A, U, $\leqslant$ and σ be as above and let $\leqslant'$, σ' the part-relation and sum relative to the Us in the role of new atoms. Fix two old atoms i, $j\eta A$. For any x, $y\eta A$, the sums $\sigma(x, i)$, $\sigma(y, j)$ are individuals of U. We can define the ordered pair (x, y) as $\sigma'(\sigma(x, i), \sigma(y, j))$. So we get the definition of function from A to A.

Since A was an arbitrary plurality, we obtain the general definition of mereological function.

8.9 Conclusions

In this chapter we have shown that—provided a machinery for pairing taken as primitive—mereology is adequate to a reformulation of megethology in first order logic with identity, without the use of plural quantifcation.

Besides, we have proved that, ordered pairs can be restricted to atoms and codified by means of finitely many pluralities.

In Chap. 7 we have proved that, in turn, plural quantification is adequate to a reformulation of megethology without the help of mereology. The two results show that Lewis' framework, endorsing both mereology and plural quantification, is overwhelming, stressing the power of each of the two components separately.

Finally, we have reconsidered Lewis' combined use of mereology and plural quantification MPQ. We have shown that, adopting a relativistic notion of atom, according to which any individual can play the role of atom, one can reconstruct Cantor's theorem and prove that, given any infinite, MPQ guarantees the existence of large and large infinities. This result seems in conflict with the alleged ontological innocence of MPQ.

References

Burgess, J. P., Hazen, A., & Lewis, D. (1991). Appendix on pairing. In D. Lewis (Ed.), *Parts of classes* (pp. 121–149). Blackwell.

Feferman, S., & Hellman, G. (1995). Predicative foundations of arithmetic. *Journal of Philosophical Logic, 24*(1), 1–17.

Fraenkel, A., Bar-Hillel, Y., & Levy, A. (1973). *Foundations of set theory*. North-Holland.

Hellman, G. (1989). *Mathematics without numbers: Towards a modal-structural interpretation*. Oxford University Press.

Hellman, G. (2002). Maximality vs. extendability: Reflections on structuralism and set theory. *Reading natural philosophy: Essays in the history and philosophy of science and mathematics* (pp. 335–361). Open Court.

Kelley, J. L., & Stone, M. (1955). *General topology* (vol. 233). van Nostrand.

Lewis, D. (1991). *Parts of classes*. Blackwell.

Chapter 9
Final Ruminations

Abstract In this final brief chapter, we aim to summarize the theses and results presented and achieved throughout this book. We offer further insights into the crucial role of imagination in logical and mathematical thought. In particular, we emphasize the significance of our acts of choice as a powerful tool for combining potential and actual infinities.

We hope that the reader, at the end of this book, has realized the crucial importance of arbitrary reference in logical and mathematical thought.

In order to make this notion perspicuous we have introduced an imaginary ideal team of infinitely many agents with the job of performing a simultaneous act of choice ad libitum.

Imagination has always been for the mathematician a precious source of inspiration for the construction of mathematical theories. However, within classical mathematics, imagination has not been adequately exploited at theoretical level.

On the contrary, intuitionism has intensively exploited the idea of free choice by developing the well-known theory of choice sequences, in particular of lawless sequences. These are constructed step by step in the course of time through a sequence of infinitely many acts of choice of an ideal agent. So they are intensional entities whose extensions are forever *in fieri*. They are in agreement with the Brouwerian conception of an indeterminate potential infinite incompatible with classical logic.

In this book we have tried to show, especially in our revision of megethology, that our theory of choices is in agreement with classical logic and supplies a way of combining the potential infinite of acts of choice with the actual infinite of classical mathematical entities.

What about mereology? Lewis maintains that mereology is ontologically innocent. In particular, he argues that, unlike set-theoretical membership, the existence of the mereological sum of any plurality X of individuals does not commit to the existence of any new individual beyond the Xs. On the other hand, he clearly treats the sum of the Xs as an outright individual, different (in general) from all Xs. In contrast, he does not regard a plurality X as a singular individual (any talk of X

M. Carrara, E. Martino, *Arbitrary Reference in Logic and Mathematics*, Synthese Library 490, https://doi.org/10.1007/978-3-031-66452-6_9

as a single entity is to be understood as a mere linguistic device to talk of the *Xs* plurally). How can it be that this single thing is nothing different from the many?

Of course, Lewis' claim, literally understood, would be contradictory. We think that it is to be understood in the sense that whoever accepts the existence of the *Xs*, commits oneself automatically to the existence of their sum.

One could maintain that the existence of the latter is merely potential. But since, according to Lewis, all subpluralities of a given plurality are thought of as simultaneously existent, so should be all their sums. It follows that the mere acceptance of an actual countable infinite forces one to accept an actual uncountable infinite as well, against the alleged ontological innocence of MPQ.

Furthermore, starting from the observation that mereology does not exploit the internal structure of atoms but only their behavior with respect to a certain order relation (called part relation), we have proposed a relativistic notion of atom. According to this, any individual, quite independently of its nature, can play the role of atom. Within such relativistic perspective, by using only the axioms of MPQ, we have reconstructed a mereological version of Cantor's theorem and shown how MPQ leads from any infinite plurality to the proliferation of higher and higher infinities.

We conclude that, quite independently of Lewis' axioms on the size of the mereological universe, MPQ, far from being ontological innocent, turns out to be a very powerful ground for a reconstruction of the mathematical infinite.

Author Index

M. Carrara, E. Martino, *Arbitrary Reference in Logic and Mathematics*,
Synthese Library 490, https://doi.org/10.1007/978-3-031-66452-6

Subject Index

M. Carrara, E. Martino, *Arbitrary Reference in Logic and Mathematics*, Synthese Library 490, https://doi.org/10.1007/978-3-031-66452-6

The manufacturer's authorised representative in the EU is Springer Nature Customer Service Centre GmbH, Europaplatz 3, 69115 Heidelberg, Germany. If you have any concerns regarding our products, please contact ProductSafety@springernature.com

Printed and bound by CPI Group (UK) Ltd, Croydon, CR0 4YY
28/09/2026
02224070-0012